THE IDENTIFICATION OF SLAGS FROM ARCHAEOLOGICAL SITES

by

Hans-Gert Bachmann

Occasional Publication No. 6

Published by the Institute of Archaeology,

31-34 GORDON SQUARE, LONDON WC1H 0PY

1982

ISBN 0-905853-10-5
ISSN 0141-8505

Published by the Institute of Archaeology,
31-34 Gordon Square, London WC1H 0PY

Printed in Great Britain by

Whitstable Litho Ltd.,
Whitstable,
Kent.

Contents

Introduction

This is NOT a monograph on slags for the extractive metallurgist, mineralogist, etc. Though emphasis is laid on slags as artificial mineral assemblages - with some of the unavoidable terminology specific to this field - the primary aim of this paper is to introduce archaeologists to finds of technological significance and to stress the aspects essential for a close association between archaeology and early technology, in particular archaeometallurgy.

The suggestion to write this introductory paper on a subject not readily in the minds of archaeologists was made in the course of lectures on archaeometallurgy. There is no need to point out the role of metals in cultural development. However, in order to manufacture metal artefacts, metals had first to be made. The many by no means primitive methods of smelting and melting are at present under vigorous examination by a number of specialists. Archaeometallurgy is just another of those interdisciplinary sciences in the general field of archaeometry, bringing together archaeologists and scientists. It can be approached from either side: the archaeological, and the scientific or technological. The archaeologist, in particular the excavator in the field, no matter how broad-minded and versatile, cannot be an expert in every discipline brought into relationship with modern archaeology. The scientist, on the other hand, called in for assistance and aid, often has difficulties in explaining his line of thought and reasoning to somebody outside his particular enclave. To make himself understood he has to omit and 'water-down' certain facts, laws and theories. The attempt to explain an important field of early technology to archaeologists takes some courage. The author, anticipating criticism from his fellow scientists, will try to familiarise, no matter how superficially, the reader with his special field of interest, and with the practical experience he has gained over the years through close and fruitful joint ventures with archaeologists.

The paper is divided into chapters which will gradually lead the way from field evidence to laboratory investigation. The many often complex and sophisticated methods will only be touched upon briefly. The true domains of the specialist should be left to the scientist familiar with them. The archaeologist is again called onto the scene when it comes to the time to draw conclusions from analytical results. His help and knowledge are essential once implications and consequences are drawn, culminating in a synopsis of one kind or another. The first part will lay the foundations and the second part will illustrate these by a series of examples. Once slags, the abundant, omnipresent, worthless, discarded 'artefacts' of early technology, are seen and appreciated in their proper context, this modest attempt - in spite of its omissions - will be justified. Hopefully the reader will realise how many questions still remain unanswered and that only by interdisciplinary co-operation may some of the problems eventually be solved.

Part I: Slags in General

INTRODUCING SLAGS

In almost every survey or excavation, the field archaeologist will encounter finds which he may tentatively label 'slags' or 'vitrified material'. In very general terms, slags are waste, ie. discarded or left behind evidences of human activity, difficult to classify and to date, but remarkably resistant to weathering. Do slags reveal anything but the fact that fire and heat were somehow involved in their making? Is there a difference at all between 'man-made' slags and the molten material (lava, obsidian, etc.) from volcanism? Are the answers to these questions relevant to archaeological research at large or is the identification of slags just another topic to which undue importance is paid by a trespassing natural scientist?

The effect of heat on inorganic material, whether accidental or intentional, will initiate reactions: the firing of clay, the burning of top soil in a domestic hearth, or the smelting of ores in a furnace. These reactions need not necessarily result in slag formation. The degree of heating plus refractory properties of the material exposed to higher temperatures very often result in products, which - though they may have suffered transformations - have not reached the molten state; they do not belong to the type of material classified as slags. A slag in its true sense is a once molten silicate or silicate mixture, sometimes including oxides, phosphates, borates, sulphides, carbides, pure metals, etc. Some of these admixtures may not have reached their melting points during the formation of the bulk slag material. They are embedded in a slag matrix like raisins in a cake. A slag can represent a simple compound consisting of just one single slag mineral or phase after solidification, but more often slags are complex mixtures of many constituents with a wide range of composition.

In the course of archaeological field work, one has to decide whether slags found at a site justify further investigation or whether they are just 'finds' hardly worth recording. For instance, a mud brick building reinforced with timber, burnt down by invaders or by accident may turn into a mass of slag, sometimes preserving the original wall structure of the building to a considerable height. Two other groups of non-metallurgical slag-like materials come to mind. There are the vitrified forts of the Celtic Iron Age, whose glass compositions are reported (with a good bibliography) by Youngblood *et al.* (1978), and the cinder mounds of the South Indian Neolithic, believed to be burned cow dung, *cf.* Zeuner (1959) and Allchin (1963). What information can be deduced from the analyses of such slags? Probably very little, except perhaps an indication of the temperature of the fire that caused the destruction, as derived from the formation temperatures of the slag minerals. On the other hand, small fragments of inconspicuous slag adhering to a crucible sherd discovered in a workshop area are likely to tell us something about the metals or alloys used at the site in question and may therefore be of paramount significance. How can one distinguish between significant and insignificant slag material? What specimens should be selected for further investigation by a chemist, and what - after all - is the archaeological implication of the laboratory results? These are only a few of the questions underlying the co-operation between archaeology and archaeometry. In order to avoid the expansion of excavation reports with endless and not always very informative tables of analytical data, some clarification and classification is necessary. In doing so, the chemist is fully aware that ancient slags are not always well-defined, homogeneous materials, easily placed in particular categories, nor can he do without the terminology and the reference to physico-chemical principles when explaining his results to the archaeologist.

However, archaeologists are becoming more and more conscious of the facilities modern analytical methods can offer. If slags are submitted for identification and analysis, the archaeologist should supply the necessary information about the site, any accompanying material, stratification, etc., and - very important - the amount of slag found. The joint venture between archaeologist and chemist requires some understanding of the possibilities and limitations offered by physico-chemical methods. None of these is a panacea to all questions. Neither does the combined application of numerous methods always give the answers the archaeologist is expecting. The following chapters are intended as a brief outline of presently available methods and their significance to archaeological research. As the archaeologist normally is not so much concerned with the practical side of analytical work, but rather interested in the results and their interpretation, importance is given more to the latter than the former.

FIELD IDENTIFICATION OF SLAGS

In an attempt to classify the various possible origins of ancient slags, the following sources have to be considered:

Geological material ('natural' slags):

- lava, pumice, obsidian (frequent in volcanic areas)
- tektites (generally rare, but quite abundant in certain regions; resembling obsidian)

Domestic and destruction slags:

- destruction of buildings by fire (eg. molten and vitrified mud brick walls), vitrified forts
- slagged structures and remains of household ovens and heating installations (eg. slagged fuel ashes, bricks, linings, etc.)

Technological, non-metallurgical slags:

- brick and tile manufacturing
- lime kilns
- pottery kilns
- glass-making and glazed stone production (eg. in early dynastic Egypt)

Metallurgical slags:

- non-ferrous metal smelting (eg. copper, lead, tin)
- non-ferrous metal melting (eg. purification of raw metals, alloying)
- bloomery hearths for wrought iron production
- iron smithing

This brief and incomplete summary leaves out slags from modern processes, like blast furnace slags. A full treatment of this is given in *Slag Atlas*, published recently by the Verein Deutscher Eisenhüttenleute (1981). According to Needham (1958) cast iron was made in China in blast furnaces at least as early as the Warring States period in the 4th century BC. There are also a number of probably accidental examples of cast iron from the Roman period, as reported by Coghlan (1956). This means that blast furnace slag phases might occasionally be found on Roman bloomery sites. Though not mentioned specifically, the summary above includes all sorts of material somehow affected by fire and heat, such as partly molten or vitrified material (cinder, dross, etc.), as well as glassy and crystalline slags proper. Perhaps one day all these relics of 'pyrotechnology' will be equally significant to archaeological research. If, for instance, fission track dating is applicable to all glasses, whether they are of geological or human origin, glassy slags may become important objects for dating. Rather high uranium concentrations (perhaps 1000 ppm) would be required for the fission track dating of materials as young as slags, although the presence of unabsorbed zircon inclusions might make the process feasible. The same is true of other dating techniques, like thermoluminescent (TL) and archaeomagnetic methods. TL dating of slags has not been a conspicuous success so far, probably because of the high concentration of transition metal ions which they usually contain, but attempts to use the most common slag mineral, fayalite, for TL dating are promising (Wagner, 1981). For the time being, slag identification has preference over dating. Learning more about slags may initiate future developments in this field. In limiting ourselves to methods which are at present available, slag identification will essentially be treated as a criterion for distinguishing between slags and 'non-slags', and further between metallurgical and non-metallurgical slags.

As will be shown in this chapter, identification should start in the field. The 'macroscopic' description of a sample as found is not only a prelude to further 'microscopic' treatment; very often, a thorough field identification makes the more elaborate laboratory work superfluous.

Contrary to the case of glass and pottery making, where broken remains and fragments of the finished product have nearly always been left behind at the production site, slags are very often the only witnesses of archaeometallurgical processes that once took place. In antiquity, non-ferrous metals were precious. Great care was taken to recover and salvage even the smallest bits and pieces from the byproducts of smelting and melting. Further, in metal production, slagging was an essential part of the pyrometallurgical process. This is the reason why metal slags serve as process indicators. Metallurgical slags are not just unwanted waste; they were made intentionally and their identification is more important than that of slags from other sources.

Slags vary in colour, appearance, texture, specific gravity, etc. Fortunately, nearly all ancient metallurgical slags are ferrous silicates, no matter whether they are the byproducts of iron or non-ferrous metal production. This facilitates field identification and distinction from non-metallurgical slags. In order to identify a find, tentatively classified as slag or slagged material, the following external properties and circumstances of occurrence should be checked:

Colour:

Due to their high content of iron silicates, slags from metal smelting or melting tend to be black or dark grey on freshly exposed surfaces. An iron tool, preferably a geologists' hammer or the like, should be used to chip off a fragment from the slag piece to be examined.

Streak:

Streak is the colour of a powdered specimen. This identification method, well known in mineralogy, is equally applicable to artificial minerals, ie. slags. The most common method of determining streak is to draw the fresh surface of the specimen across a plate of white, unglazed porcelain, called a streak plate. As a substitute, the protruding, unglazed rim of the reverse side of a saucer, cup or plate can be used. Ferrous slags give a dark brown to black streak, while, for instance, slagged mud bricks - even if they are dark grey externally and may therefore be mistaken for metallurgical slags - only give a very faintly brown or uncoloured streak. However, some metallurgical slags are extremely hard and do not powder readily.

Texture:

Many slags show flow textures, often resembling volcanic lava. Other samples are striated or layered. Glassy slags will reveal conchoidal fractures, while crystalline slags exhibit reflections or lustre from crystals (needles or laths) of slag minerals.

Porosity:

Though a great number of naturally occurring minerals resemble slags, a look at the fresh surface, preferably with a magnifying lens (magnification of 5 to 8 times is sufficient) will nearly always answer the question: mineral or slag? Unlike most minerals, slags show gas holes ('gassiness'). They vary in size from minute cavities to large circular or elliptical vesicles. Because some dark volcanic rocks, like basalts, are also occasionally porous, the geology of the site where the slags were found should be taken into consideration. The Roman preference for millstones hewn out of black or dark grey basalt lava, often transported over long distances, may sometimes be puzzling. Experience will help to overcome such occasional obstacles. A word of warning: slag identification requires close visual examination. The discovery of dark spots or layers during a survey, seen from the distance, can lead to wrong conclusions if closer inspection of the dark material is omitted. Surface layers of dark brown or nearly black sedimentary formations sometimes found in fragments in the detritus of desert valleys (wadis) have in the past been mistaken for slag accumulations, while they are in fact scattered blocks of sandstone layers coloured by iron and manganese oxides. Close examination of such 'pseudo'-slags will immediately reveal the lack of pores and gas holes. In glassy slags, pores are sometimes missing, but very rarely are these the only type of slag material found at a particular site.

Inclusions:

Very few slags are free from inclusions like mineral grains, fragments of charcoal, globules or veinlets of pure metals or metal sulphides (ie. matte), sand grains from the slag bed, and the like. Identification of these inclusions again requires a fresh unweathered surface, preferably one that was made with a rock cutter's saw. A magnifying lens and a small bottle (glass stopper!) with hydrochloric acid (15%) will be of good service. The acid will help to distinguish between carbonate and quartz, etc. inclusions. Carbonates (calcite and related minerals, sometimes formed as secondary minerals after exposure of the slag to weathering for lengthy periods) will give off carbon dioxide with effervescence.

Rust:

Due to the high iron content of early slags, light to dark brown rusty surfaces are quite frequent. On the other hand, many slag minerals, especially iron silicates, are resistant towards weathering, even under humid climates, and retain their original colour. If small or medium-sized rusty lumps are found during an excavation or field survey, a close examination is always justified.

Smell:

Slags from the smelting of sulphidic ores very often have a characteristic smell of sulphur compounds, even if present only in small amounts. This property can become quite important in the later course of examination as an indication of the type of ores used at the smelting site where these slags were found.

Specific gravity:

Most metallurgical slags are rather heavy (specific gravity 3 to 5 g/cm^3) and crystalline (needles or laths visible on a fresh surface), but some material is - due to rapid cooling - glass-like and resembles obsidian, a naturally occurring volcanic glass. As stated above, it is not very likely that an ancient slag heap or layer consists entirely of glassy slags.

Metallurgical indications:

A piece of black, heavy slag, tentatively identified as a metallurgical slag, normally does not tell one anything about the kind of metal with which it is connected. The only exception is the occasional presence of green spots of secondary copper minerals on slag surfaces, formed by the weathering of metal inclusions in the slag. This is proof of the slag's origin from copper smelting. If such colour indicators are missing, only chemical analysis either in the laboratory or by field tests (blowpipe, portable XRF, etc.) can tell whether iron, lead, copper or tin have been smelted or worked at the site. Somewhat easier is the distinction between tapped and furnace slags. Tapped material normally shows the obvious signs of flow texture, ie. smooth surface with runners, elongated ridges or strings of formerly more or less viscous material that has flown out of a tap-hole or the like. The same features apply to crucible slags, if they have been poured. Materials not exhibiting these signs may tentatively be classified as furnace slags, pending further examination. Smithing slags tend to have a plano-convex or concave-convex shape; their low silica and high iron-oxide content make them particularly susceptible to weathering. They are brittle and very often coated with light to medium brown crusts of iron oxides and hydroxides.

Amount of slag found at a site:

Archaeometallurgical sites vary in the amount of slag they have produced. This may range from a few isolated stray pieces to thousands of tons. Apart from slag properties, such as weight, colour, porosity, inclusions, etc., it is important to estimate the amount of material found at a particular site. Already a single smelting process in a small furnace or hearth will produce a sizeable heap of waste. Isolated pieces of slag may indicate smithing or crucible smelting/ melting in workshop areas, cf. Tylecote (1974). In order to get an idea about the type and duration of metallurgical activities that went on at a particular site, it is advisable to carry out a rough calculation of the amount of material discovered:

slag-covered area (in m^3)

x depth or thickness (in m)

x specific gravity of the slag (in g/cm^2)

x packing factor

= amount of slag (in metric tons)

Field determination of specific gravity and packing factor is somewhat difficult; average values of 3.5 g/cm^3 for specific gravity and 0.8 for packing factor resp. will give reasonably good estimates.

Sampling:

If possible, sampling should be carried out at the site by the person in charge of the subsequent analytical investigation. Experience has shown that archaeologists tend to select the most conspicuous pieces while neglecting the plain average types. Some slag analyses cited in publications are not from genuine slags but from slagged furnace linings, etc. The value of information from such data is correspondingly low. No general rules can be given about the number of samples that should be taken from a specific site. If the slag material is uniform and present in considerable amounts, a single specimen may already give important hints. However, it is usually safer to submit a series of five to twenty samples for analytical inspection, particularly if the outer appearance already indicates a variation in the nature of the material found. Conclusions drawn from the analytical results derived from a larger number of samples are much more significant and can be used for statistical evaluation. Variation in composition is often quite typical for early smelting processes, and can only be judged adequately by representative sampling. If the total amount of slag found does not exceed a few kilograms, it is always advisable to collect all of it. Apart from proper labelling, slags should be packed in strong cardboard boxes rather than in plastic bags. If paper or plastic bags are used, the slag samples should first be wrapped in newspaper, because slags tend to have sharp cutting edges.

Magnetism:

Quite often slags are magnetic. This is caused by a sometimes appreciable content of the ferromagnetic mineral magnetite, Fe_3O_4, and/or metallic iron. To test for magnetism in the field, the slag has to be crushed and powdered (in a mortar or on a piece of hard rock or steel plate). A small hand magnet (easily obtained from a toy shop) is useful to carry out this test.

Typology:

A number of authors have introduced slag typologies based on outer appearance. For copper slags Zschocke and Preuschen (1932: 99) have given the following classification for the various types of slags found at the Mitterberg, Austria, smelting sites: blocks of slag, coarse broken slag, coarse plate slag, homogeneous plate slag, and slag gravel. For iron slags, the most comprehensive list so far was compiled by Sperl (1980a); he distinguishes between the three main groups: furnace slags, tap slags and cakes of slag. Each group is divided into three or four sub-groups. Normally, a large smelting site will reveal a variety of types of slags. These should be classified and characterised by visual inspection and, if possible, the approximate ratio of the various types should be recorded and representative specimens of each type collected.

Stratigraphy:

Slag heaps, whether shallow or high, often exhibit pronounced layering. Each layer represents a production period, however short, and may serve to fix a chronology (in most cases only relative) and consequently to document the development of technological processes. However, we have to bear in mind that an industrial site where slags were deposited is often submitted to more changes in shorter intervals than a habitation site. Furnaces may be built into old slag heaps, and slags can be shifted and older slags piled on younger layers if there is need for new working space, thereby reversing the chronological sequence. A profile through a slag heap often shows all the features of what is known to sedimentologists as cross-bedding. The recording of such a profile, in very much the same manner as an archaeological profile is drawn and measured, is strongly recommended, though trenches through slag heaps pose problems of safety as the crumbly material is likely to collapse.

CHECK LIST FOR SURVEYS OF ARCHAEOMETALLURGICAL SITES

Slags have to be seen not as isolated pieces of evidence. The material accompanying them is of equal importance. Investigating archaeometallurgical sites, the author has found it useful to use a check-list to record all the essential circumstances and to avoid omitting relevant details. The characterisation of slags forms only part of this list, but the intention is obvious: to bring slags into a context of related and associated finds and circumstances.

For example, a check list could include the following items:

Characterisation of Site:

Site No.
Map/Coordinates
Name of locality/situation
Metal smelted and/or worked at site
Types of operation (smelting, refining, casting, etc.)

General Geography and Geology:

Topography of the area (including water courses, wadis, vegetation, etc.)
General geology (deposits of ore, clay, fluxes, etc.)

Raw Materials found at the Site:

Ores (with gangue)
Fluxes
Charcoal, wood
Refractory materials

Slags:

Type(s)
Amount(s)
Average size of fragments

Slags (cont.):

Average size of whole cakes (weight, thickness, etc.)
Colour
Streak
Texture
Porosity
Inclusions
Weathering
Process relation (tapping, etc.)

Furnace Materials:

Lining:

Colour
Temper
Slagging/vitrification
Curvature
Ratio to slags

Stones:

Petrological classification
Slagging/vitrification
Size
Ratio to slags

Tuyeres:

Shape/size (length)
Construction
Outer diameter
Inner diameter
Material
Temper
Slagging/vitrification
Ratio to slags (how many fragments in relation to total amount of slag found)

Crucibles:

Material
Temper
Shape (curvature, estimated volume)
Outside appearance
Inside appearance
Adhering metal, slag, etc.

Associated Finds:

Pottery
Flints
Hammer stones
Crushing plates
Other tools

Associated Structures:

Building remains
Storage pits
Wells
Walls
Roads, paths

LABORATORY IDENTIFICATION OF SLAGS

So far we have treated slags in a manner familiar to every archaeologist when dealing with finds. The description by visual inspection, ie. the 'outer' appearance and the correlation with other evidence forms the basis of all subsequent investigations and treatments.

Once the finds have reached the laboratory, the analyst will have to decide what part of the material should be submitted for further analyses. As sawing, grinding and powdering will destroy the original appearance of the specimens, portions of the samples as found should be kept - if possible - for further reference.

Chemistry and physics nowadays provide powerful methods for slag identification and characterisation. Among the standard methods in archaeometry, the following are in particular applicable to slags:

Chemical composition:

- Chemical analysis (wet-chemical, 'classical' methods)
- Emission spectroscopy (inlcuding ICP = Induced Coupled Plasma spectroscopy)
- Atomic absorption spectroscopy (AAS)
- X-ray fluorescence analysis (XRF, alone or in combination with scanning electron microscopy or microprobe analysis)
- Neutron activation analysis (only for trace element determinations)

Phase composition:

- Visible light microscopy (thin and/or polished sections under normal and polarised light)
- Electron microscopy (in particular scanning electron microscopy (SEM) and electron probe microanalysis = microprobe)
- X-ray diffraction

Physical constants:

- Specific gravity
- Melting point or 'softening' interval as determined by differential thermal analysis or hot stage microscopy
- Viscosity (difficult, but informative)

Age determinations:

- Radiocarbon dating of associated or included charcoal and - if present in sufficient quantity - of carbides in iron smelting and smithing slags
- Thermoluminescence dating, provided suitable mineral inclusions or fragments of furnace lining, crucible sherds, etc., are available
- Fission track dating (applicable only to glassy slags), provided the uranium content is sufficiently high

With the introduction of rapid modern methods to find out the chemical composition with a high degree of accuracy, including trace elements, the old-fashioned wet-chemical methods have been somewhat neglected. They are, however, still the most reliable methods to distinguish between di- and trivalent iron. As ancient slags are basically mixtures of iron compounds it is of significance for subsequent calculations and classifications to distinguish between these two valencies. Many early slags contain compounds known as spinels, ie. oxides of di- and trivalent metals like Fe, Al, Mn, etc. Among the spinels most frequently found in slags is magnetite, Fe_3O_4. If the trivalent iron contained in magnetite is not determined as such, errors will result in the course of compound calculation, eg. the theoretically calculated amount of fayalite, Fe_2SiO_4, a silicate of divalent iron, is automatically too high if ALL the iron contained in a slag is only recorded as divalent iron, despite the fact that some of it may be in the trivalent state.

Another method of determining the content of trivalent iron in a slag is by measuring its magnetism or rather its 'magnetisability'. As shown by Presslinger *et al.* (1980) this can be done quite rapidly by commercial instruments, so-called 'ferrite meters'.

A full treatment of the various methods mentioned is beyond the scope of this paper. The reader is referred to the excellent summary given by Tite (1972).

These investigation methods are not specific for ancient slags. They are part of the standard repertoire of archaeometric research and are equally useful for the analysis of other finds. With the exception of age determinations, the analytical chemist will treat ancient slags just as he would modern ones. The interpretation of the results, however, has to take into account ancient methods of extractive metallurgy, which may have little or no relation to present day processes.

The archaeologist's request for slag characterisation does not always require the application of the full range of the methods mentioned. As a rule, the identification and specification should not be limited to the determination of the chemical composition alone. The information about phases, ie. the chemical compounds of which the slag is composed, is equally important. Only the combined results of chemical and phase composition will give the necessary background for the evaluation and reconstruction of ancient metallurgical processes.

Assuming the archaeologist and the analyst have decided what methods to apply to answer specific questions, the outcome of the laboratory characterisation will have to be added to the field characterisation. Thus, we now have the macroscopic and the microscopic fingerprint of the slags with which we are dealing. In short, the following essentials are known:

Sample appearance (visual inspection)
Chemical composition (analysis expressed as oxide, etc. components in per cent)
Phase composition (mineral content as revealed under the microscope or by other methods, eg. x-ray diffraction).

Some workers stop at this point. No doubt the data accumulated so far can be made up into impressive tables, but with all this information at hand, it is desirable to proceed further. In order to do so, we have to interrupt the logical sequence aimed at a synopsis. The following chapter introduces us to the role of slags as process indicators.

FORMATION OF METALLURGICAL SLAGS

The reduction of metal compounds to pure metals without slag formation requires reagent grade chemicals and special care to avoid contamination with the material of the reaction vessel (crucible, etc.). These conditions can only be realised in a laboratory, almost never in technical processes using naturally occurring materials. The use of ores is invariably related to the formation of slags, because slags act as necessary collectors for impurities, introduced into the process by the ores, the gangue, the furnace lining and the charcoal ash. Of course, the purer the ore, the less slag will be formed. Refining processes, like crucible melting, give very little slag and sometimes none at all, perhaps only some dross or a slagging ('glazing', vitrification) of the ceramic material of the crucible used for the operation. Sometimes the ores plus associated gangue have a chemical gross composition ideal for slag formation. Such ores are called self-fluxing. This property lacking, no sufficient separation of metal and impurities can be achieved in the course of the smelting process. Early metallurgists gradually learned by experience how important it was to add the right type of flux to the furnace charge in order to produce a good slag, ie. a slag of low melting point and viscosity which could take up all the impurities. Thus, the composition and the properties of metallurgical slags are influenced by:

Ores and gangue
Fluxes added
Furnace construction material (including lining, bricks, tuyeres, temper added to clays used as furnace linings, etc.)

Charcoal and wood ashes
Process conditions (heat distribution, air intake, furnace profile and height, retention time of slags within the furnace)
Cooling conditions
Weathering (dependent on time, climate and mode of deposition).

Each piece of slag more or less displays all or part of these influences. The analytical chemist with a basic knowledge of extractive metallurgy will try to reconstruct the conditions and procedures that had to take place in order to form this particular 'process witness', ie. the slag he is investigating.

In order to obtain a slag with a low content of metal, the molten silicates should have a low viscosity, a low density and high surface tension; the latter to diminish penetration into and reaction with the furnace lining (cf. Schmiedl *et al.*, 1975).

In spite of the many complex influences that have to be considered, certain assumptions and simplifications can be made when dealing with ancient slags as compared with modern products:

- The temperatures in early smelting furnaces and hearths rarely exceeded 1300 to 1400 degrees centigrade; in most cases they were lower;

- For heating and reduction only wood and charcoal were employed, other materials being unknown. An exception is the use of coal for smelting iron in China by the 4th century AD and probably even earlier; *cf.* Needham (1958);

- The choice of fluxes was limited. Apart from the gangue of the ores, only iron and manganese oxides, lime, silica and perhaps fluorspar were used;

- For the construction and lining of furnaces material preferably from local resources was employed. High quality refractories can be ruled out in antiquity.

Keeping these limitations in mind, ancient metallurgical slags are very much like modern ferrous slags and their characterisation criteria are nearly identical.

CHEMICAL AND PHASE COMPOSITION OF SLAGS

The ferrous slags formed in bloomeries as well as in copper and lead smelting have as their major components FeO, CaO, and SiO_2; in many slags, some of the iron is also present as Fe_2O_3. In addition, the slag may contain variable amounts of MgO, MnO, Al_2O_3, P_2O_5, alkali oxides, etc. Further, the presence of metal inclusions in ancient slags is more a rule than an exception.

To study the relationship between chemical composition and compound formation during and after the solidification of slags, we have to introduce the concept of phase diagrams. By definition, a phase is any portion, including the whole, of a system which is physically homogeneous within itself and bounded by a surface so that it is mechanically separable (eg. solid - liquid, two solids in mixture, etc.) from any other portion. The evaluation and interpretation of phase diagrams is a special and rather sophisticated field of physical chemistry. It is beyond the scope of this paper, primarily intended for the guidance of archaeologists, to present this complex subject in detail. The reader is referred to standard textbooks on physical chemistry. A very concise summary of phase diagrams in general with a full glossary of the terminology can be found in Levin *et al.* (1974: 5). The few explanations necessary for a basic understanding have largely been adopted from this source.

Quite a few workers in the field of slag research have chosen ternary systems for the representation of the analytically determined composition of slags, normally given as oxide percentages (eg. SiO_2, FeO, Al_2O_3, etc.). A complete graphical representation of a ternary system is difficult to achieve if the variables: composition, temperature and pressure, must be included. However, the vapour pressure of silicate systems is so low as to be negligible. The condensed system, omitting pressure, can be represented by means of triangular co-ordinates for the composition (three components) and the additional indication of uniform temperature intervals by isotherms in the same way as elevation contours are drawn on topographical maps. Fig. 1 illustrates how the composition of a ternary system is represented by triangular co-ordinates. Each side of an equilateral triangle is divided into ten parts (one part = ten weight per cent), each division being intersected by lines parallel to each of the other two sides. The point C at the apex of the triangle is composed wholly of component C (= 100 weight per cent). Any point on the base line A-B is composed entirely of components A and B with none of C. The relative distance of point X from each of the three apices may be expressed as a percentage and thus represents the percentage composition of a ternary mixture in terms of components A, B and C. Point X, for example, has a composition of 45 per cent A, 20 per cent B and 35 per cent C.

Apart from the fact that all phase diagrams represent systems in equilibrium, only the major oxide components of slags can be included in ternary diagrams. Any graphical representation, therefore, has to be a compromise. The system CaO-FeO-SiO_2, for instance, was chosen by Lupu (1970), Milton *et al.* (1976) and Steinberg and Koucky (1974) for the representation of early copper smelting slags from Timna, Israel, and from Cyprus respectively. To avoid the neglect of other oxide components also present in the slags analysed, Lupu (1970) extended the apices of the ternary system CaO-FeO-SiO_2 to (CaO+MgO)-(FeO+MnO)-(SiO_2+Al_2O_3). The errors introduced by this expansion are not too serious for the particular type of slags concerned.

Another proposal for a ternary phase diagram for slags was suggested by Morton and Wingrove (1969b; 1972). Their fundamental research on Roman and medieval iron smelting slags (ie. bloomery slags) is based on a system which has as its corner points FeO, SiO_2 and the compound $CaAl_2Si_2O_8$, known to mineralogists as anorthite, a member of the feldspar family. Anorthite is also a very common constituent of slags, predominantly occurring as a glassy phase. This system, apparently, is ideal for the description of iron smelting slags; cf. Fig. 2. Due to the great similarity between iron and many non-ferrous metal smelting slags, this system is equally valid for the representation of copper and lead slags, provided the CaO contents are not too high. Tylecote and Boydell (1978), therefore, have used this system to classify copper slags from laboratory smelting experiments, simulating early smelting techniques from Timna. However, ancient copper slags from the numerous smelting sites in this area vary considerably in their CaO contents. To account for the wide range of composition of Timna copper slags, Bachmann (1980) has employed the two systems FeO-SiO_2-$CaAl_2Si_2O_8$ and FeO-$CaAl_2Si_2O_8$-$CaSiO_3$ ($CaSiO_3$ = wollastonite). The first named system is reserved for slags low in CaO and the second for CaO-rich slags.

All the ternary systems mentioned are part of the more general quarternary system (CaO+BaO+Alk_2O)-(FeO+MnO+ZnO+MgO)-Al_2O_3-(SiO_2+P_2O_5). The analyst will have to decide from case to case whether the whole system or a ternary section of it comes closest to an adequate coverage of the slag samples he wants to classify. Sometimes he may have to introduce additional systems, like the system FeO-Fe_2O_3-SiO_2, to treat his data accordingly.

Despite necessary simplifications, which have to be made when attempting to plot chemical analyses of slags in phase diagrams, a wealth of information can be deduced from these diagrams in the way of thermodynamics, crystallisation sequences, activity of components for compound formation and the like. As shown in modern text-books on extractive metallurgy, for example, Rosenqvist (1974: 324), the full treatment of metallurgical processes, no matter whether they are ancient or modern, is incomplete without the implications of phase relations and properties as provided by phase diagrams, established by many workers for systems of pure components under controlled conditions. The progress made in archaeometallurgy during recent years was made possible by using the interpretative possibilities offered by this branch of physical chemistry. Though the archaeologist is primarily interested in the identification of slags, a basic understanding will help to judge reports on archaeometallurgical research, particularly those dealing with the reconstruction of ancient processes. Identification techniques can be generalised to some extent to cover the majority of possible types of slags. Interpretation, however, has to treat sites and sometimes even single samples individually. The author's aim is a presentation of identification methods; therefore, interpretation will only be touched upon briefly.

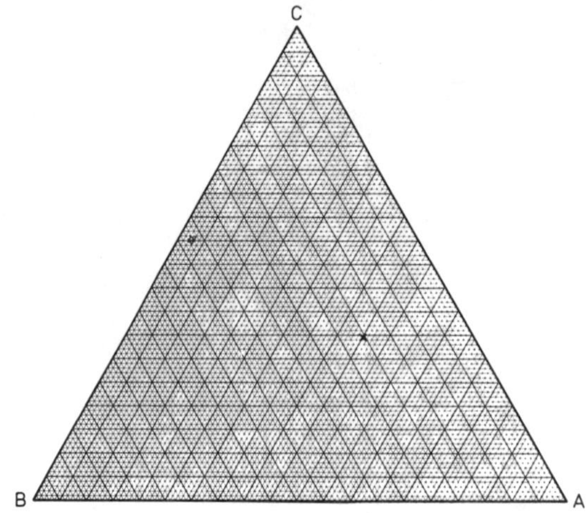

Fig. 1

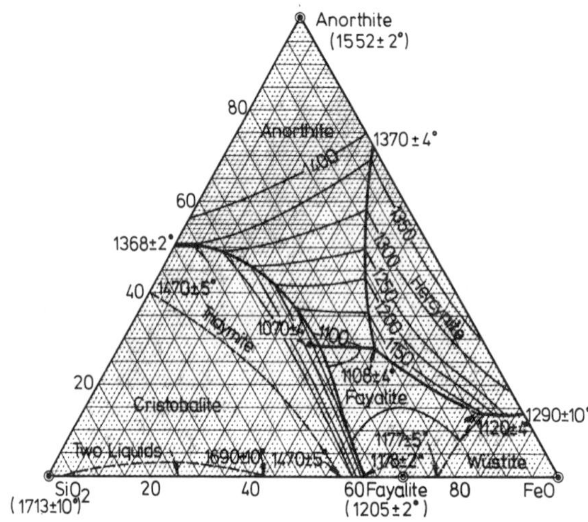

Fig. 2

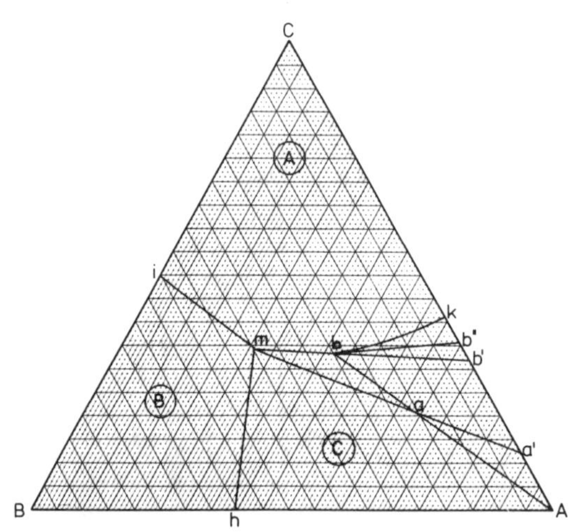

Fig. 3

To give the reader an idea about the deductions that can be drawn from ternary phase diagrams, we shall follow what is called the 'crystallisation path' in a simple theoretical ternary system adopted from Levin *et al.* (1974). A knowledge of the crystallisation curve or the melting curve (the reverse of the crystallisation curve) for any particular melt is very valuable in the study of phase formation during the solidification of slags. In Fig. 3, point *m* is the 'ternary eutectic', ie. the point at which all the crystallisation curves of this system terminate. If a slag melt of composition *a* is chosen and allowed to cool, the system remains liquid until the 'liquidus' is reached, the more or less curved plane separating the liquid from the solid, at which temperature the solid *A* begins to crystallise. The course of the crystallisation from this point to the boundary *m-k* follows a straight line drawn through *A* and *a*. Due to crystallisation of component *A*, the melt alters its composition. It is 'loosing' component *A* which starts to crystallise. As the liquid melt changes in composition from *a* to *b*, more and more *A* crystallises. At *b*, a second phase (mineral) appears and the crystallisation curve follows boundary *k-m* (a 'valley route' in the 'landscape' of our ternary phase 'map') with phases *A* and *C* crystallising together. At point *m*, the temperature remains constant with slag minerals *A*, *B* and *C* crystallising together until the remainder of the molten slag has solidified. The final product will be a mixture of large crystals of *A* and *C*, and small crystals ('eutectic mixture') of *A*, *B* and *C*. The composition of the solid slag minerals crystallising at any instant during the process of cooling along *b-m* is given at the point where the tangent to the crystallisation curve intersects that side of the composition triangle representing the two solid phases co-existing. For example, at *b* it is indicated by the intersection of the tangent to the curve *m-b-k* at the point *b* with the line *A-C* at point *b*". The ratio of *A* to *C* is given by the ratio *b*"-*C*/*b*"-*A*. The mean composition of the two solid phases that have crystallised between points *b* and *m* is represented by the intersection of a line drawn through *m* and *b* and the side of the composition triangle at *b*'. In this case it is a mixture of *A* and *C* in the proportion *b*'-*C*/*b*'-*A*. The mean composition of the total solid slag which separates out between *a* and *m* (before *B* begins to solidify) is determined by drawing a line through *m* and *a* to the side of the composition triangle at *a*'. During eutectic crystallisation at *m*, the composition of the total solids, ie. all the portions of the slag having crystallised during the continuous fall in temperature, changes from *a*' to *a*, reaching the latter point as the last drop of the melt disappears.

SLAG MINERALS

Slags are basically silicates. Admixtures of other compounds (oxides, metals, etc.) are more often present than absent. The various identification techniques applied to ancient slags have in the past revealed a large number of individual phases or minerals in slags. The term 'mineral' in its strict sense is only valid for naturally formed compounds, while in slags we are dealing with artificial, man-made material. The use of the term 'mineral' in slag identification is justified, because artificial minerals in slags nearly always resemble their naturally occurring equivalents. As mineral names are used widely in reports on slags, a brief catalogue of the main compounds contained in slags will be given. This list is not complete. To avoid confusion, the rare and less important components are omitted from this list.

Silica and silicates:

The main silicates in ferrous slags can be classified according to the ratio between metal oxides (MeO) and silica (SiO_2):

ratio $MeO:SiO_2$	silicate examples
2 : 1	fayalite = $2(\underline{Fe},Zn,Mg,Mn)O.SiO_2$
2 : 1	monticellite = $(\underline{Fe},Zn,Mg,Mn)O.CaO.SiO_2$
1.5 : 1	åkermanite = $(\underline{Fe},Zn,Mg,Mn)O.2CaO.2SiO_2$
1 : 1	pyroxene = $(\underline{Fe},Zn,Mg,Mn)O.CaO.2SiO_2$

- Quartz and its high temperature modification cristobalite, SiO_2, are only present in slags of relatively high silica content together with fayalite, pyroxene, etc. SiO_2-phases in slags are frequent though anomalous; if they occur together with oxides, like magnetite and other spinels, silicate forming reactions were prevented by lack of temperature (ie. reaction energy), interruption of the smelting process, etc.

- Fayalite = Fe_2SiO_4 or $2FeO.SiO_2$ is the only real compound existing in the binary system $FeO-SiO_2$. The phase $FeSiO_3$ or $FeO.SiO_2$, a theoretical Fe-pyroxene, is unstable. However, in SiO_2-rich slags with sufficiently high amounts of MgO and/or CaO, the FeO-rich pyroxenes hypersthene, $(Fe,Mg)_2Si_2O_6$, or hedenbergite, $CaFeSi_2O_6$, can be formed. Fayalite may take up a maximum of 9 per cent of ZnO. ZnO contents above this limit will give rise to the formation of willemite, Zn_2SiO_4. Isomorphous replacement of Fe by Mn is frequent; Mn_2SiO_4 = tephroite, $(Mn,Fe)_2SiO_4$ = knebelite. Also small percentages of CaO can be incorporated in fayalite. With higher CaO contents, minerals of the monticellite group will be formed; $MgCaSiO_4$ = monticellite or $FeCaSiO_4$ = Fe-monticellite or $MnCaSiO_4$ = glaucochroite. A complete series of mixed crystals exists between Fe-monticellite and fayalite. Fayalite in all its varieties is one of the most common minerals in ferrous slags.

- Willemite = Zn_2SiO_4 or $2ZnO.SiO_2$, very often with additions of FeO, has been observed in lead smelting slags when using ZnO-rich ores. Idiomorphous willemite crystals, occurring as laths with hexagonal cross-sections, are frequently associated with ZnO-rich fayalites. The crystallisation of this mineral takes place prior to that of fayalite. Willemite crystals are free from inclusions; they can take up 20 per cent or more of FeO.

- Monticellite = $(\underline{Fe},Mg,Mn)CaSiO_4$ or $(\underline{Fe},Mg,Mn)O.CaO.SiO_2$ is more correctly written with the general formula $RO.CaO.SiO_2$, with R = Fe,Mn,Mg,Zn. The mineral occurs in slags with lower SiO_2 and higher CaO content, eg. in some lead slags. Fe-Zn-monticellites are not infrequent. Glaucochroite, the Mn variety, $CaMnSiO_4$, has not yet been observed in ancient slags, but can certainly be formed if sufficient MnO and CaO are available. There is complete solid miscibility between fayalite and monticellite, but in nature as well as in slags only fayalites with up to 20 per cent of monticellite (and vice versa) are known.

- Pyroxene = $(\underline{Fe},Mn,Mg)O.CaO.2SiO_2$ or $(\underline{Fe},Mn,Mg)CaSi_2O_6$ is quite common in metal slags. The naturally occurring alkaline pyroxenes are missing in slags. Their genesis requires Al_2O_3; this, however, is normally present in slags only in minor amounts. Pyroxenes are very rarely the only silicate phases present; according to the chemical composition of a particular slag, pyroxenes are in most cases associated with orthosilicates, like fayalite. The compound corresponding to the pyroxene mineral hedenbergite, $CaFeSi_2O_6$, a rock and slag forming mineral, is not part of the ternary system $CaO-FeO-SiO_2$. It originates from wollastonite-type mixed crystals, $(Ca,Fe)O.SiO_2$, in the course of a solid state reaction at about 950 degrees centigrade.

- Wollastonite = $CaSiO_3$ or $CaO.SiO_2$ can take up 67 per cent of $FeO.SiO_2$ and/or equivalent amounts of $MnO.SiO_2$. The mineral vogtite with the formula $(Ca,Fe,Mn)O.SiO_2$ is isomorphous with wollastonite. If present in slags, it will probably undergo transformation to pyroxenes just like Fe-rich wollastonites.

- Rhodonite = $(Mn,Fe,Ca)SiO_3$ or $(Mn,Fe,Ca)O.SiO_2$, a mineral similar in its chemical compositions and crystallographic properties to wollastonite, has only been observed in modern acid (silica-rich) Bessemer slags, not yet in ancient slags.

- Melilites are silicates with the general formula $(Ca,Na,K)_2(Mg,Fe^{2+},Fe^{3+},Al)(Si,Al)_2O_7$. They are solid solutions of the following end members:

$Ca_2Al_2SiO_7$	=	gehlenite
$Ca_2MgSi_2O_7$	=	åkermanite
$Ca_2Fe^{2+}Si_2O_7$	=	ferroåkermanite
$Ca_2Fe^{3+}SiAlO_7$	=	iron gehlenite
$Ca_2ZnSi_2O_7$	=	hardystonite
$NaCaAlSi_2O_7$	=	sodium melilite
$KCaAlSi_2O_7$	=	potassium melilite

With the exception of potassium melilite, miscibility among the various end members is complete. They are frequently encountered in blast furnace slags and other smelting slags of non-ferrous metals, but also in slagged furnace linings (chamotte, etc.).

- Feldspars, though the most common silicates in nature, play only a minor role in slags. Due to the low alkali content of most slags, alkali feldspars have only rarely been observed in metallurgical materials. Plagioclase feldspars of the anorthite type, $CaO.Al_2O_3.2SiO_2$ or $CaAl_2Si_2O_8$, are sometimes present in iron, etc. slags. In slags with a sufficiently high Ba content celsian, $BaAl_2Si_2O_8$, may occur.

- Micas, like biotite, $K(Mg,Fe,Mn)_3(OH,F)_2AlSi_3O_{10}$, are very rare constituents of slags; their formation depends on a certain amount of fluoride being present in the smelting charge from fluorspar, etc.

- Cordierite, $Mg_2Al_4Si_5O_{18}$, is not a mineral occurring in ferrous metallurgical slags. It is only included in this list because it is the most important phase in the system $MgO-Al_2O_3-SiO_2$, a system in which many non-metallurgical slags, technological or accidental ones, can be accommodated. Therefore, cordierites, including the iron-variety of this mineral, $Fe_2Al_4Si_5O_{18}$, together with feldspars, iron oxides, glasses, etc., have been observed in slagged or partly melted argillaceous material, like mud bricks, fused furnace linings from pottery, etc., kilns.

- Glasses in metal slags are considered by Morton and Wingrove (1969b; 1972) to be generally of anorthite-like composition. Other glasses in ferrous slags with a composition of 47 per cent FeO and 35 per cent SiO_2 in a molar ratio of 1:1.12 correspond in their composition to pyroxene and fayalite. Glasses are formed if the molten slag is cooled rapidly. Glass formation is also dependent on the composition of the melt. Lack of certain ingredients promoting crystallisation may play a part just as the non-equilibrium of certain processes. Lead in lead slags is to a great extent present as lead silicates; however, so far no definite crystalline lead silicate phases have been observed in these slags. Therefore, we must assume that lead silicates preferably tend to form glasses. This is not surprising, considering the importance of lead silicates in ceramic glazes, which are also glasses.

Oxides:

Next to silicates, oxides are the most important constituents of slags; nearly all of them are iron oxides.

- Wüstite, 'FeO', is a common mineral in bloomery slags, but it is also present in many ferrous copper and lead slags.

- Magnetite, Fe_3O_4, is an iron oxide containing both di- and trivalent iron. Divalent iron can be partly substituted by Mg, Mn and Zn, and trivalent iron by Al and Cr. Magnetite is present in nearly all metallurgical slags with medium to high FeO content. This mineral is either embedded in the silicate matrix of the slag in the form of small grains or crystal skeletons, or in idiomorphous crystals, sometimes of considerable size. Magnetite is stable even in the presence of sulphides. Magnetite is a member of the spinel family of oxides and is frequently associated with other spinels.

- Spinels, apart from the iron spinel magnetite, are oxides of di- and trivalent elements such as Al (trivalent), Zn, Mn, Mg and Fe (divalent). Pure end members of spinels are rare. Their trend to form mixed crystals is evident from the zonal growth of spinel crystals, with each zone having a different composition of slightly varying ratios of tri- and divalent metal ions. Spinels are among the first phases to crystallise from a melt. Skeletons of octahedra or sharply defined distorted octahedra are normally embedded in the silicate matrix in the same manner as magnetites. Depending on the Al content of the melt, iron spinels (ie. magnetites or ferrites) or aluminium spinels (ie. spinels proper) are formed; eg. Zn in lead slags is known to occur either as a ferrite or as a spinel.

- Hematite, α-Fe_2O_3, may be present as segregations in magnetites if furnace conditions had been highly oxidising. This oxide of trivalent iron is an oxidation product of magnetite, and not as common a slag constituent as magnetite and other spinels.

- Delafossite, $CuFeO_2$, is sometimes found in melting slags (eg. crucible slags); its presence in smelting slags is less pronounced but possible. The mineral, often exhibiting zonal growth patterns, crystallises in plates, sometimes curved (visible in cross-sections under the micrscope), or in laths, occasionally surrounding magnetite crystals as an outer rim. Delafossite segregations in magnetites as well as eutectoid intergrowth of delafossite with cuprite, Cu_2O, have been observed.

- Cuprite, Cu_2O, requires for its formation partly oxidising conditions; it is therefore more frequently found in melting (crucible) slags than in smelting slags. In polished sections, the mineral appears blue to grey with dark red inner reflections. It either occurs in large irregular masses or in small grains, often surrounding metallic copper as a rim.

Sulphides:

Sulphides in slags are to be expected if sulphidic ores had been used. No matter what type of process is employed to convert these ores to metal, the sulphur has to be removed. The oxidation to sulphur dioxide - the primary mode of removal - is seldom complete. Relics of sulphides can thus become entrapped in slags. Matte, a technical term in extractive metallurgy given to partly roasted, ie. oxidised, sulphidic ores, after melting to a sulphide or sulphide/oxide mixture, is not a slag, but an intermediate product which is further processed. Though the identification of matte is a topic separate from that of slag identification, sulphide inclusions in slags have nevertheless to be regarded as slag minerals.

- Copper sulphides, frequently associated with iron sulphides in slags, are an indication that sulphidic ores were smelted. As only small amounts of copper sulphides are soluble in slag silicates, the sulphides are present as a separate phase or phases. Inclusions of mixed copper-iron sulphides, ie. matte, in slags often resemble metallic copper in their appearance. They have the colour of dark brass or bronze. The most common copper and copper-iron sulphides encountered in copper as well as in some lead slags are:

covellite = CuS

chalcocite = Cu_2S

bornite = Cu_5FeS_4

chalcopyrite = $CuFeS_2$

cubanite = $CuFe_2S_3$

These minerals often occur intergrown with each other. Droplets of sulphides may contain magnetite inclusions.

- Iron sulphides, especially pyrrhotine, FeS, are found in lead and sometimes copper slags. Very often, eutectoid intergrowth with magnetite, together with chalcocite, Cu_2S, and bornite, Cu_5FeS_4, is observed. These sulphides often form small nests in slag silicates. Matte tends to be a collector for various elements, eg. precious metals. The different distribution of certain elements between matte, slag and copper metal was recently investigated by Yazawa (1980). This line of research is worth following up also for ancient slags. It could prove to be interesting for studies of provenance.

- Lead sulphides are very rare in slags. They have occasionally been found in slags with very high PbO contents.

- Zinc sulphides are restricted to slags with high Zn and S contents. They are very seldom found in ancient slags. Occasionally, lead slags from the famous Attic mines of Laurion contain a few per cent of zinc from the smelting of lead-zinc ores. Zinc sulphides may vary in appearance. Small drops and rods intergrown with silicates are just as possible as crystal skeletons or irregular masses, commonly intergrown with small or even minute crystals of magnetite. Occasionally the two crystallographic modifications of ZnS, sphalerite and wurtzite, both of which can take up Fe and Mn to some extent, have been identified in lead slags.

Antimonides and arsenides:

The many minerals belonging to this group, very often of complex composition, are only likely to occur in ancient slags if so-called 'fahlerz', ie. ores with appreciable amounts of metal antimonides and/or arsenides, were smelted. The highly sophisticated smelting techniques necessary to treat these ores economically without loosing too much metal in the slags, were not known in antiquity. If such fahlerz was used accidentally, the results must have been disappointing (cf. section on speiss).

Metals:

Though it is the intention of every metallurgist, from early times on, to keep the content of pure metals in the slag as low as possible, one had and still has to compromise; no slag is entirely free of metallic inclusions; only the amount varies from antiquity to modern times.

- Lead in its well-crystallised metallic form (cubes) is quite a typical constituent of ancient Greek lead slags from the mines and smelters of Laurion.

- Copper metal as droplets, prills or globules, varying in size from fractions of a millimetre up to several millimetres, is found in most early copper slags. If not visible to the naked eye, a microscopic examination will almost certainly show metal inclusions, usually in rounded shapes, but rarely also as minute veins. Metallic copper inclusions serve as an indication whether a ferrous slag is related to iron or copper smelting.

- Iron, more specifically wrought iron, made by the bloomery process is obtained as spongy masses, isolated nodules, flaky particles, etc., in solid form. Only modern blast furnaces are able to produce liquid iron; cf. Needham (1958) and Coghlan (1956) (p.2 above). The temperatures that could be reached in bloomery hearths and furnaces were normally too low to approach the melting point of iron, 1535 degrees centigrade. Metallic iron, therefore, is very common in bloomery slags and to some extent in smithing slags, but is not normally to be expected in copper and lead slags. However, in copper slags with stoichiometrically unbalanced iron and silica contents (ie. more iron oxides in the change than can combine with silica to form iron silicates like fayalite, etc.), metallic iron may form under strongly reducing conditions. Iron droplets thus formed will normally segregate through the liquid slag if its viscosity is sufficiently low. At the furnace bottom, the iron particles will be taken up by the accumulated molten copper. During solidification at the end of the smelting run, the two metals will separate, because iron is practically insoluble in solid copper. This explains the sometimes considerable contents of metallic iron in early copper ingots, clearly visible in polished sections. Correspondingly, small iron particles may also be found in copper smelting slags, generally in the slag layer immediately above the copper ingot. Nothing has as yet been reported about metallic iron inclusions in early lead smelting slags.

SECONDARY SLAG MINERALS

A slag may undergo more or less extensive weathering depending on composition, climate and duration of deposition. Sulphides (matte, etc.) are converted to sulphates. Soluble sulphates are aggressive and can react with other slag components to form a variety of further sulphates, of which only insoluble ones, like gypsum, $CaSO_4.2H_2O$, tend to remain in the slag thus leached. Beautiful crystals of gypsum often fill vesicles in slag. In addition, leaching may alter the overall composition of the slag significantly. Chemical analyses of such material have therefore to be interpreted with due care.

Ancient copper slags nearly always show green spots and incrustation on their outside. These are copper chlorides, sulphates and carbonates which were formed after deposition.

Slags containing iron oxides are often covered with a weathering crust of brown iron hydroxides. Also iron sulphates are frequently encountered, especially on slags resulting from the smelting of sulphidic ores. These are but a few examples of secondary mineral formations. Some more will be given in the second part of this paper.

To the author's knowledge. up until now no general treatment of the mineral content of ancient slags has been published. Information may only be found in occasional papers dealing with individual sites or projects. For this reason, the list of minerals given is necessarily incomplete. With increasing knowledge in the field of archaeometallurgy, the list certainly will have to be expanded. As it stands, it may serve as a first tentative guide.

For modern metallurgical slags, the situation is of course different. Metallurgists have long since realised the importance of slag mineral identification for process control. Though it may often be misleading to draw parallels between ancient and modern slags, published sources on modern material by Trojer (1974: 393, 1963), Obenauer (1954: 459) and Faber (1954: 519) were consulted in order to compile the list given.

ADDITIONAL SLAG PROPERTIES

The various criteria obtained by the methods outlined above are in most cases sufficient to characterise a slag adequately, but in order to extract the maximum of information from slags, metallurgists are likely to ask for further properties. Physical properties like melting point, viscosity and density have already been mentioned. Determination of surface tension is sometimes required to determine the 'wetting' of metals and refractories by slag. It may become significant also for ancient slags with the developing progress of archaeometallurgy. Some properties, like electrical conductivity, have only to be considered in modern electrometallurgical processes. Other slag properties likely to be of importance for the extractive metallurgist are the slag basicity, the thermodynamic behaviour and the content of gases in slags. The thermodynamic properties of slags are a function of the free energy of mixing and the chemical activities of the oxide components, like SiO_2, CaO, FeO, etc., in slag mineral formation. The thermodynamics of ferrous slags to which the majority of ancient metallurgical slags belong have been studied in detail, however, in temperature ranges not usually applicable to ancient slags. It depends to a large extent on the partial pressure of oxygen of the slag systems under consideration; *cf.* Rosenquist (1974). In modern industrial practice, slag basicity is commonly expressed by some basicity index. Different expressions have been suggested to relate the weight percentages or molar fractions of 'basic' oxides, like CaO, MgO, FeO, etc., to those of the 'acid' oxides SiO_2 and Al_2O_3. Ratios like

$$\frac{CaO}{SiO_2} \quad \text{or} \quad \frac{CaO + MgO + FeO \text{ etc.}}{SiO_2 + Al_2O_3}$$

serve the purpose of describing the relative basicity of a slag, because an absolute measure of slag basicity does not exist. A method of calculating slag basicity from analytical data, applied to ancient slags, in order to predict the formation of certain slag minerals, has been suggested by Bachmann (1980).

The ratio of the 'basic' oxides against 'acid' oxides, expressed in weight percentages, as given above, is also used to formulate a 'viscosity index'. The oxides CaO, MgO, FeO, MnO, Alk_2O tend to lower the slag viscosity, while SiO_2 and Al_2O_3 increase it. Thus, the ratio of the two sums:

$$K_v = \frac{\Sigma CaO + MgO + FeO + MnO + Alk_2O}{\Sigma SiO_2 + Al_2O_3} \quad \text{(in weight percentages)}$$

may vary between 0.5 and 1.0, approximately. The lower the viscosity index, the higher the viscosity. Of course, viscosity is largely dependent on the temperature of the slag in question. But, in order to achieve a low viscosity - a property of paramount importance for the separation of metal from slag - the slag must have a composition which with rising temperature will result in a sufficiently low viscosity. As ancient smelters could not increase the temperature in their furnaces beyond an upper limit of about 1300 to 1400 degrees centigrade, the composition of slags - influencing the parameters melting point and viscosity - had to be optimised by using appropriate fluxes. Fortunately, low melting fayalite slags have a high viscosity index (ie. low viscosity) as well. They account for the efficiency of some of the ancient smelting processes known to us.

The attention given to slag properties by metallurgists is documented by the many papers and text-books on this subject. Together with other data connected with metal smelting and refining, like furnace construction, gas composition, etc., process performance can be predicted and guided. Contrary to the task of the modern metallurgist to optimise technical processes with a maximum of available data, the archaeometallurgist has to be content with limited and scanty information on processes very often not yet fully understood. He will hardly ever be able to gather all the evidence and data necessary for a complete reconstruction of ancient processes. He just has to be content with what he finds and to extract as much as possible from these chance relics. For some aspects of his problems there may well be an abundance of information, while for others he will encounter gaps only to be filled with assumptions and hypotheses. As said at the beginning, slags are the most abundant materials related to metallurgy that have survived time and destruction. Treating slags in what may seem at first sight a rather exaggerated way has to be regarded as a precautionary measure, in order not to miss a single item which eventually may prove helpful in putting together the great puzzle of ancient technology. We are still at a level of knowledge where data accumulation, ie. analysis in its widest sense, is necessary. Some sort of a foundation has to be laid first before we can proceed to synthesis, ie. the reproduction of ancient smelting techniques and practices by experiment. A fair number of iron smelting experiments have already been undertaken and we have a reasonably good understanding of the bloomery process. Also, on copper smelting practices, experimental evidence has been collected and further work is in progress at a number of laboratories. Eventually, more will be made known on ancient lead production, silver cupellation and tin smelting.

Part II: Catalogue of Selected Examples

The following part is essentially a catalogue of various types of slags. One way to introduce the uninitiated to the problems of slag identification is through a set of pictures. A look into the 'microcosmos' of slags with the aid of polished sections of slags under the microscope, or the vistas of slag morphology opened by the scanning electron microscope, can tell more than mere words. There is no such object as THE typical copper or iron slag, just as there are no single, simple processes producing slags. Each piece is an entity in itself. However, there are general trends and laws of formation, though we do not yet know all of these. For many of the examples illustrated in the catalogue, chemical composition was determined as well. These analytical data are summarised in Table 1.

NON-METALLURGICAL SLAGS (Plate I)

Evans and Tylecote (1967) have given a number of examples of some vitrified products of non-metallurgical significance. The material they analysed, all light in colour and of low density, included glass slag, vitrified wood ash, vitrified sea weed ash, vitrified cow dung, and burnt haystack. Glass slags and vitrified products tend to be high in alkali. The lime content of wood ashes often exceeds 40%, whereas that in glass is much lower. The ratio of K/Na in wood ash is much greater than one, while it is considerably less than one in seaweed ash due to more sodium being taken up by the sea water; see also Biek (1970).

True metallurgical slags, ie. those of approximately fayalitic composition, nearly all exhibit a fairly high iron and/or manganese content, they are low in alkali and normally also low in phosphorus (<3% P_2O_5). However, metallurgical slags are occasionally 'contaminated' with furnace lining, and in extreme cases these slagged linings approach non-metallurgical slags in composition. To illustrate this group of material, we have chosen fused mud bricks as well as slagged furnace linings from two copper smelting sites in the Timna area, southern Israel.

Plate I shows partly fused and slagged mud bricks from the Assyrian trading colony of Kaneš (Kültepe) in Central Anatolia (Turkey), dating from the 15th century BC:

Ia) light brown, very porous

Ib) dark brown, relatively dense.

(For chemical compositions, see Table 1.)

COPPER SLAG IN *STATU NASCENDI* (Plate II)

An attempt to smelt oxidic copper ore nodules (malachite, etc.) in a bowl hearth has resulted in what will be called a furnace 'conglomerate', found in Wadi Amram, southern Israel, either of very early date or a much later trial by nomads to 're-invent' copper smelting. In any case, it is an example of how it all could have started.

IIa) A cut-off slice of bowl hearth filling with partly decomposed and fused ore fragments, droplets of copper metal, lumps of charcoal and signs of the very beginning of slag formation, ie. molten silicates filling interstices;

IIb) Same as a)

COPPER SLAGS FROM SMELTING OXIDIC ORES (Plates III and IV)

The smelting process involves reactions, which can be summarised as follows:

Decomposition of Ore:

$$CuCO_3.Cu(OH)_2 \xrightarrow{<500^oC} 2CuO + CO_2 + H_2O$$

Reduction:

$$2CuO + CO \longrightarrow Cu_2O + CO_2$$

$$Cu_2O + CO \longrightarrow 2Cu + CO_2$$

Slag Formation:

$$SiO_2 + 2FeO \xrightarrow{>600^oC} 2FeO.SiO_2$$

Either iron oxide or silica can act as flux, depending on the gangue of the ore (*cf.* Tylecote, 1979).

Plates III and IV represent typical examples of copper smelting slags from the Timna area, southern Israel, probably the best known ancient metallurgical centre today. The main activities there date back to the 10th and 12th centuries BC. Locally mined oxidic ores were smelted at numerous individual sites, leaving behind slag heaps varying in quantity from a few tons to thousands of tons. A full range of analytical data for these slags is given by Bachmann (1980). A typical example of the composition of a copper slag from smelting oxidic ores with the addition of manganese oxides as flux is given in Table 1. Further information on the smelting of oxidic ores can be found in Bachmann and Rothenberg (1980) and Bachmann (1978).

IIIa) Fayalite crystals, fractured and partly dissolved in a glassy matrix (x50);

IIIb) Same as IIIa), but with additional small particles of metallic copper (white) (x200);

IIIc) Dendrites (fir-tree like agglomerates of tiny crystals) of iron oxides (wüstite and magnetite; white), associated with fayalite (light grey), glassy matrix (dark grey) and vesicles filled with resin (from preparation of section; black) (x200);

IVa) Large fractured crystals of fayalite (grey) plus gas 'holes', ie. vesicles (black) (x150);

IVb) Same as IVa); large fractured and partly dissolved ('corroded') fayalite crystals, typical of primary crystallisation, ie. large crystals with idiomorphic (= well-developed) crystal faces, embedded in a matrix (dark grey) containing silicates from secondary crystallisation, probably also fayalites (grey) (x360);

IVc) Dendrites of iron oxides (wüstite and magnetite; white) in fayalitic copper slag (x360);

IVd) Spheres of metallic copper (white) plus dendrites of iron oxides (light grey) (x360);

IVe) Same as IVd) (x360);

IVf) Scanning electron microscope (SEM) picture of a cluster of well-developed fayalite crystals adhering to the wall of a vesicle (gas hole) (x3000);

IVg) Fragment of cellular charcoal (acacia?), embedded in slag (x90).

COPPER SLAGS FROM SMELTING SULPHIDIC ORES WITH MATTE, ETC. (Plates V to XIII)

In a schematic summary, the following process steps have to be associated with the production of copper, matte and slag using sulphidic ores:

Dead Roasting:

$$\underset{\text{(chalcopyrite)}}{CuFeS_2} + 3O_2 \longrightarrow FeO + CuO + 2SO_2$$

Partial Roasting:

$$4CuFeS_2 + 7O_2 \longrightarrow 2Fe_2O_3 + 4CuS + 4SO_2$$

Matting:

$$CuFeS_2 + 5O_2 + \underset{\text{(pyrite)}}{2FeS_2} \longrightarrow \underbrace{FeS + CuS}_{\text{(matte)}} + 2FeO + 4SO_2$$

Matte Roasting:

$$CuS + FeS + 3O_2 \longrightarrow CuO + FeO + 2SO_2$$

Reduction:

$$CuO + CO \longrightarrow Cu + CO_2$$

Slagging:

$$2FeO + SiO_2 \longrightarrow 2FeO.SiO_2$$

Refining:

$$2Fe\ (+\ Sb,\ As,\ etc.) + O_2 \longrightarrow 2FeO\ (+\ impurities)$$

(cf. Tyelecote, 1979).

The smelting of sulphidic ores is thus a much more complex process than smelting oxidic ores. The slags from these ores show a great variety, often exhibiting the particular stage of the process to which they are related. The examples chosen come from a variety of sites.

Compared with slags from smelting oxidic ores, those from sulphide smelting are more abundant, though, of course, this depends on the available ores. While obviously the first copper ores ever submitted to reduction to copper metal must have been the spectacular green secondary ores, Bronze Age copper technology had already succeeded in smelting copper sulphides. Thus, the majority of copper slags found world-wide belongs to this group. Actually, the slags from smelting oxidic ores can be regarded as a subgroup of copper slags in general. Both types are similar. Only the former lack the inclusion of sulphides (like matte, relics of primary ore particles, often in a state of decomposition). The mineral content of copper slags is similar to that of lead slags with the silica content normally somewhat higher in copper than in lead slags. The copper content is present either in the form of metallic droplets, copper oxides or sulphides, the latter being practically insoluble in the silicates. Contrary to modern blast furnace slags compared with iron bloomery slags, ancient copper slags resemble their modern counterparts in many respects. Among the minerals found (in ancient as well as in modern material) is magnetite (often with inclusions of metallic Cu). Magnetite can form during the smelting process in the following manner:

$$Cu_2O + 3FeO \longrightarrow Fe_3O_4 + 2Cu$$

Magnetite occurs either in fairly large crystals or as skeletons. Al spinels are rare, as generally the Al content of slags is too low. Among the silicates, fayalites are predominant, often associated with pyroxenes. Melilites are rare in copper slags, as the CaO content is normally too low to form this mineral. If melilites are present, their crystals are often surrounded by an outer rim of pyroxene, of presumably hedenbergite-type. Very frequent, and in considerable quantity, is a glass component or matrix of varying composition, generally specified as anorthitic. Among the many possible copper sulphides likely to occur are neodigenite, $Cu_{2-x}S$, chalcopyrite $CuFeS_2$, etc. Delafossite, $CuFeO_2$, is another quite common slag component. It often has a zonal structure and tends to exhibit a eutectoid pattern consisting of Cu_2O and $CuFeO_2$. Cu_2O, cuprite, if present, appears in convoluted aggregates.

Va) Partly decomposed (probably by roasting) sulphidic copper ore fragment (chalcopyrite) from Alaca Hüyük, Northern Anatolia (Turkey), 16th century BC (by radiocarbon dating); sample no. 1 (x200);

Vb) Same as Va), eutectoid decomposition pattern with copper sulphides (white), magnetite (medium grey) and wüstite (light grey) (x200);

Vc) Same as Vb) (x500).

Sample no. 2 from the same locality and of the same age is from a similar piece of partly decomposed primary sulphidic ore.

VIa) Mixture of iron oxides (grey), plus fractured and partly decomposed particles of presumably chalcopyrite (white), interspersed with gas holes (black) (x100);

VIb) Dendrites of iron oxides (grey), penetrating copper sulphides (x100);

VIc) Same as VIb) (x100).

A completely different piece of slag from the same locality and of the same age is depicted in Plate VII; it is a thin homogeneous plate slag which was tapped (sample no. 3):

VIIa) Dendrites of wüstite and magnetite embedded in a matrix of fayalite and glass (x50);

VIIb) Same as VIIa) (x200);

VIIc) Same as VIIa) and VIIb) (x200).

The following three slag samples from Kitzbühel, Austria, are similar in appearance and composition; for chemical composition see Table 1. They are typical tap slags from matte smelting in low shaft furnaces known as 'Krummofen' and in operation in the Tyrol until 1860. The blue-black slags with occasional green weathering dots on their surfaces resemble iron smelting slags from the bloomery process, with the exception that bright particles show up frequently which have to be interpreted as matte, metallic copper and cuprite, Cu_2O; eutectoid intergrowth is often exhibited. The copper content goes up to 4%, indicating low process efficiency. As the sulphur contents were not determined analytically, it is difficult to tell the ratio between metallic and sulphidic copper. The amount of metallic iron is appreciable and that of trivalent iron low, both pointing to strong reducing conditions and/or insufficient addition of silica flux. The age of the slags is presumably medieval.

VIIIa) Upper surface of sample; scale 0.7;

VIIIb) Front view of sample; scale 0.7;

VIIIc) Dark grey fayalite crystals with faint contours in a matrix, together with scattered minute to medium sized bright particles of matte and iron oxides (x200);

VIIId) Isolated particles of a complex mixture of multi-phase matte, embedded in slag matrix; a spinel (magnetite?) crystal is visible at the upper rim of the matte inclusion (x500);

VIIIe) Same as VIIId) under polarised light, showing the different phases of the matte particle more clearly (x500);

VIIIf) Same as VIIIe), but with the polarising filter in the oblique position (x500).

Another piece of material from Kitzbühel (sample no. 7) is presented in Plate IX:

IXa) Bottom side of sample; scale 0.9;

IXb) Side view of sample; original size;

IXc) Section through sample (x2), showing clearly the scattering of light grey to white matte particles throughout a silicate matrix (medium grey);

IXd) Iron oxides (light grey) and matte particles (bright) in dark grey silicate matrix with fayalite crystals (x200);

IXe) A piece of metallic copper, crossed by veinlets of copper sulphide and intergrown with copper oxide (cuprite) (also in colour; x240);

IXf) A droplet of copper sulphides (matte), composed of several individual crystals, surrounded by almost white iron oxide (magnetite) crystals, plus fayalite crystals (medium grey) in matrix (dark) (also in colour; x600).

The third specimen from Kitzbühel (sample no. 8) shows similarities as well as differences compared with the other two slags. The three pieces give an idea of the variations in apparently identical samples.

Xa) Bottom side of sample (x1.5);

Xb) Side view of sample (x1.5);

Xc) Section through sample (x2), with a distribution of white to light grey metal and matte particles in a silicate matrix;

Xd) White metallic copper and matte droplets, light grey iron oxide in crystals and oval-shaped masses, surrounded by large fayalite crystals and a cryptocrystalline matrix (x200);

Xe) The centre particle (white) is metallic copper, the light grey veinlets have to be interpreted as copper sulphides and the medium grey component as cuprite. The surrounding matrix exhibits iron oxides (magnetite and wüstite) and dark grey fayalite (x240; also in colour: sulphides: blue; cuprite: red; and metal: bright yellow);

Xf) A metallic copper droplet is penetrated and surrounded by copper sulphides. In the matrix can be seen smaller spheres of sulphides and metal plus cubic crystals of iron oxides (x240; also in colour);

Xg) A metal particle with a complex intergrowth of copper oxide (cuprite) and sulphides under higher magnification (x600; in colour).

With the following Plate we return to Bronze Age copper smelting. The site, Hirschwang-Prein, Austria, has been described by Hampl (1953) and Mayrhofen (1953). Hampl noticed at a number of sites in the area a variety of material and evidence, including coarse slag, cakes of slag, thick and thin plate slags, tuyeres, and pot bellows, as well as matte, blister copper, charcoal, fluxes (quartz, iron oxides) and furnace linings. The sample illustrated on Plate XI is obviously heterogeneous coarse slag.

XIa) A particle of decomposed chalcopyrite (white) in a matrix of silicates, containing fragments of quartz (dark grey) (x34);

XIb) Same as XIa). Among the several phases present in the sulphide particle, well-developed cubic crystals of spinels (magnetite?) can be distinguished (x335);

XIc) Another section of the same sample showing decomposed sulphides (chalcopyrite and covellite) surrounded by fayalite (x34);

XId) Same as XIc) (in colour); fayalite (dark grey), chalcopyrite (white), covellite and other related copper sulphides (blue and violet) (x100);

XIe) Same as XIc); eutectoid mixture of various mineral phases, like magnetite (in the shape of light grey dendrites) and covellite (medium grey) inside partly decomposed chalcopyrite, surrounded by fayalite (dark grey) (x135);

XIf) Same as XIc); eutectoid mixture of chalcopyrite plus magnetite (?) (white) and covellite (blue) and veinlets of cuprite (red) intersecting covellite (in colour; x500).

In 1967 C. Böhne (1968) attempted to reproduce ancient copper smelting techniques using chalcopyrite. After roasting selected pieces of rich ore, smelting was successfully carried out in a low shaft furnace, with a charge of partly roasted ore to which powdered lime and fluorspar were added as fluxes. The analyses of ore, matte and slags from this smelting experiment are given in Table 1. A piece of slag, ie. slag from smelting roasted matte, was investigated, and gave the photographs illustrated in Plate XII:

XIIa) Lattice of orientated fayalite needles (light grey) in a glassy matrix, plus an isolated spinel crystal (dark to black) and droplets of metallic copper (x500);

XIIb) Same as XIIa) in colour. The spinel crystal appears ruby red (probably $FeO.Al_2O_3$) and the metallic copper whitish red (x500).

The microscopic results were verified by X-ray diffraction and X-ray fluorescence. The major phases are fayalite and metallic copper. Apart from iron, silica and copper as main constituents, the material contains some nickel (0.1 to 0.5%).

Bronze Age Cypriot copper slags from Hala Sultan Tekke and Kalopsidha, cf. Bachmann (1974, 1976), were identified using X-ray diffraction, scanning electron microscopy, and XRF (cf. Table 1). Among the components found in the slags originating from smelting sulphidic ore are: copper, cuprite, wüstite, fayalite, nantokite, CuCl, paratacamite, $Cu_2(OH)_3Cl$, $CuFe_2O_4$ (spinel phase), matte, carphosiderite, $Fe_3(SO_4)_2(OH)_5.2H_2O$ and magnetite. The pictures on Plate XIII are a selection of the SEM photographs taken from these samples:

XIIIa) Cavity in slag filled with gypsum crystals of secondary origin, ie. product of weathering and leaching after deposition (x45);

XIIIb) Coating of magnetite crystals (octahedra) on a fissure crack of tap slag (x170);

XIIIc) Same as XIIIb) (x850);

XIIId) Vesicle (from degassing) in slag, upper right hand corner: globular mass of iron sulphate (carphosiderite), due to surface weathering after deposition (x80);

XIIIe) Same as XIIId) (x800);

XIIIf) A cluster of fayalite needles at the rim of a degassing hole inside tap slag (x1600).

On problems of smelting sulphidic copper ores and related slags the following recent papers have to be mentioned: Presslinger *et al.* (1980), Rostoker and Sadowski (1980) and Sperl (1980b).

COPPER SLAGS WITH TIN, ETC.

Some copper slags contain tin in amounts from 0.1 to 0.5%. The number of tin-containing copper slags may even be larger, but very often tin is not looked for by the respective analysts. When smelting copper ores with noticeable tin contents, reduction of tin is favoured in the presence of copper due to intermetallic affinity (Maréchal, 1962). Therefore, the manufacture of low tin bronzes was possible at a much earlier time than the reduction of tin ores to tin metal. Otto and Witter (1952: 28) analysed tin-containing copper slags from Central Germany (cf. Table 1). They were derived from smelting mixed tin-copper ores, ie. tin-rich vein ores with baryte as gangue. The specimens are from excavations, but of unknown age. The sample from Opitz is claimed to come from a 'prehistoric' smelting site. Slag No. 2 from Ranis contained 0.62% of metallic copper, which itself was alloyed with 1.2% of tin. This is considered to be proof that during the reduction of the tin-containing copper ores much of the tin alloys with the copper to form a low-tin bronze.

The characterisation of copper smelting slags from smelting mixed ores (chalcopyrite plus sphalerite plus galena etc.) as they occur, for instance, in the Hartz mountains, North-west Germany, would justify a further chapter. These ores and their slags have certainly played a major role in medieval metallurgy. Their possible importance also in earlier times is still under investigation. The slags, chemically distinguishable from other copper slags by significant amounts of lead and zinc (in addition to the usual constituents) are the subject of a series of papers by Laub (1970/71; 1975; 1978); a typical example is given in Table 1.

TIN SLAGS

Despite the importance of tin as an alloying element for copper, tin slags are rare and only a few have been analysed. A summary of tin slag analyses was given by Tylecote, in Miles (1975). Among the material cited is an early tin slag from Portugal. It was found at an Iron Age hillfort dating from the 8th to 2nd centuries BC. Its composition was:

Fe_2O_3	64.46%
SiO_2	21.46%
Al_2O_3	9.34%
MnO	0.40%
TiO_2	0.96%
P_2O_5	0.30%
S	0.005%
Cu	tr.
SnO_2	2.03%

The internal structure of the slag shows transition from a grey-black matrix with some metallic tin, to fayalite laths with wüstite or magnetite. The spherical globules of tin were about 150 to 200µ m in diameter. Some $FeSn_2$ phase was also present. Hematite or limonite may have been used as fluxes. However, in Cornish tinstones, according to Tylecote, *l.c.*, iron is the main impurity. This ore would need fluxing with silica (sand). The smelting temperatures were estimated to be up to 1300°C.

In a recent paper by Tylecote (1980: 1) the analysis is given of a piece of unstratified black tin slag from the Cornish tin smelter of Calenick, which was one of the longest-running tin smelters in Cornwall (1702-1891) and one of the first to introduce reverbatory furnaces. Though this is a fairly modern slag, its accurate analysis is worth quoting:

FeO	32.60%	
Fe_2O_3	3.90%	
SiO_2	30.00%	
CaO	1.60%	
Al_2O_3	11.10%	
Mn	0.15%	
P	0.02%	
S	0.55%	
Sn	7.90%	
Cu	1.48%	
Pb	0.02%	
Sb	0.01%	
As	0.01%	
rest	10.48%	(H_2O, O_2)

Another analysis of a Cornish tin slag of unknown age is included in Table 1.

High tin content in tin slags is by no means abnormal. Cornish slags give values varying from 0.8 to 2.8% and show evidence of metallic tin. A slag from Dartmoor is said to contain 13.5% Sn, mostly as oxides, according to Tylecote (1980). In the analysis of the Calenick slag the iron content was introduced by iron oxides used as flux; copper and sulphur are from copper sulphides associated with the tin ore (tinstone). Some sulphur may have come from the culm coal used for smelting. The tinstone contained 45 to 63.7% Sn (pure SnO_2 has 79% Sn); the gangue was mostly silica. Tin smelting has been going on in Cornwall for 4000 years, but its evidence on the ground is not at all obvious. This is probably true of other tin smelting sites as well; *cf.* Tylecote (1966).

LEAD SLAGS (Plate XIV)

Lead ores are easy to reduce. Though complete slagging of the impurities requires temperatures above 1000° centigrade, the melting point of metallic lead is low (328° C) and any metal reduced from its ores easily separates itself from the rest of the furnace charge. Thus, the heat of a camp fire is sufficient to smelt lead, though very inefficient. The most important lead ore, galena, PbS, requires conversion from sulphide to sulphate prior to reduction. This is achieved either by a separate roasting step or in the upper oxidising region of a shaft furnace. During smelting a number of reactions will take place, such as:

$$PbSO_4 + 4CO \longrightarrow PbS + 4CO_2$$

$$PbSO_4 + Pb \longrightarrow 2PbO + SO_2$$

$$PbSO_4 + PbS \longrightarrow 2Pb + 2SO_2$$

$$3PbSO_4 + PbS \longrightarrow 4PbO + 4SO_2$$

$$2PbO + PbS \longrightarrow 3Pb + SO_2$$

Some lead ores contain other elements as well (like Zn and Cu). They too may enter the slags. The zinc that has not volatised during the process can form sulphides (sphalerite and wurtzite) or combine with trivalent metals to spinels; some may be present in silicates (fayalite, monticellite, hedenbergite and melilite). High Zn contents may give rise to the formation of willemite, Zn_2SiO_4. Iron, as in all the other slags described, will form silicates and oxides (mainly magnetite); very rarely iron sulphides can be detected (either in zinc sulphides or as separate iron sulphides). Copper, if present, is mostly found as sulphides like covellite, bornite, chalcopyrite and cubanite, very rarely 'entrapped' in zinc sulphide and magnetite. The presence of copper in the silicate phases is only in traces. The lead slags very often contain appreciable amounts of lead, which are retained as lead silicates but more often as metallic lead. A recent study of primitive lead smelting was made by Hetherington (1980).

One of the most famous lead mines in antiquity was situated at the southern tip of Attika, Greece, at Laurion. The Laurion mines were worked extensively by the Athenians at the time of Pericles for their content of silver. The ore, mainly galena, was largely smelted on the seashore, and the slags were dumped into the sea. There are a number of minerals specific to slags from Laurion, such as

laurionite	$Pb(OH)Cl$
paralaurionite	$Pb(OH)Cl$
penfieldite	$Pb_2(OH)Cl_3$
fiedlerite	$Pb_3(OH)Cl_4$
phosgenite	$PbCO_3.PbCl_2$

found in lead slags that have reacted with the sea water. Among the minerals thus formed are the ones mentioned above, plus cerussite, anglesite, and other secondary lead minerals; paralaurionite is the most common associate; cf. Dana (1951: 62).

Plate XIV shows 'scenes', taken by SEM, of a piece of ancient Laurion slag lying in sea water off the coast for many centuries. The lead-rich tap slag has undergone chemical changes through reaction with sea water, including the formation of new minerals.

XIVa) Lead crystals; surface-etched by sea water, on pyroxene crystals (x1500);

XIVb) Same as XIVa) (x2000);

XIVc) Same as XIVb) (x3000);

XIVd) Phosgenite crystals inside tap slag (x335);

XIVe) Paralaurionite (?) crystals (x1500).

The chemical analysis of the sample shown in Plate XIV is given in Table 1.

SPEISS (Plate XV)

Many ancient lead slags - and occasionally also copper slags - show inclusions resembling metal. Metallurgists have given these 'pseudometals' the name 'speiss'. They are a product obtained by smelting complex ores (eg. 'fahlerz'). As the specific gravity of speiss tends to be even higher than that of matte, it collects underneath the matte, or more often - as its quantity is usually small - it is dispersed in the slag. Chemically, speiss is a compound of metals (iron, copper) with arsenic and/or antimony. Speiss only occurs as a separate phase if As and Sb are present in considerable quantities. Small amounts of these elements can easily be taken up by the matte. However, with increasing copper content of the matte, its ability to absorb As and Sb decreases. Speisses could not be processed further with ancient methods of extractive metallurgy. Even nowadays they present serious problems to the extractive metallurgist. Two typical compositions of speisses are given below:

	Cu	Ag	Pb	Bi	As	Sb	Fe	Co	Ni	Zn	Sn
Cu-rich speiss	51.7	.175	35.2	1.6	2.7	3.3	1.6	0.2	0.3	1.8	1.4
Fe-rich speiss	25.8	.085	16.7	-	11.0	13.5	22.7	1.1	1.6	3.3	4.1

These samples, found together with ancient lead slag, come from Gümüşköy ("silver village") near Çiftehan, County Ulukişla in the province of Niğde, southern Turkey, about 40 kilometres to the west of the Cilician Gates at the Northern slope of the Taurus Mountains (Bolkar dağlari). Only a few kilometres to the south of the village of Gümüşköy is situated the famous deposit of Bolkarmaden (Bulgurmaden during the Ottoman Empire), a lead zinc deposit rich in silver and gold. The ore veins are of hydrothermal origin. Modern mining started in the 19th century. After a period of high activity during the second half of the last century, the mine was abandoned in 1892. Extensive slag heaps are reported near Gümüşköy and the neighbouring village of Madenköy ("mines' village"). Dating is difficult, as the mining and smelting commenced in Hittite times (19th to 12th century BC) and continued well into the Byzantine era.

This speiss sample is illustrated in Place XV. The chemical analysis of a lead-rich specimen gave the following results:

Fe	65.30%
Pb	3.60%
S	2.53%
Cu	0.76%
Mn	0.27%
Zn	0.11%
C	0.09%
P	0.07%
Si	0.06%
As	rest (>15%)

A lead free sample investigated by X-ray diffraction contained the following phases: α-Fe, $AsFe_2$ and α-FeOOH(?), the latter as weathering product.

XVa) Crystals of α-Fe embedded in $AsFe_2$ (x10);

XVb) α-Fe crystals (white) together with an eutectoid (grey) of iron and $AsFe_2$; etched with alcoholic picric acid (x80);

XVc) Same as XVb) (x80);

XVd) Same as XVb) (x400);

XVe) Same as XVb) (x400);

XVf) Eutectoid of Fe and $AsFe_2$ with droplets of metallic lead in centre of picture; etched with alcoholic nitric acid (x80);

XVg) Same as XVf), but etched with sodium picrate (x80);

XVh) Same as XVf), under normal light, lead droplets appear medium grey (x400);

XVi) Same as XVh), but under polarised light; (matrix dark and lead medium grey) (x400).

IRON SLAGS FROM SMELTING AND SMITHING (Plates XVI-XXVII)

These are by far the most common slags the archaeologist is likely to encounter. As this monograph is confined to archaeometallurgy - though this rule has not always been strictly obeyed - the field of blast furnace slags is omitted; cf. p.2. More than the smelting of non-ferrous metals, the bloomery processes give a variety of waste material which cannot just be called slag. Apart from the slags as such, cinders and bears have to be considered.

The following definitions were suggested by Morton and Wingrove (1969a): "Slag, cinder and bear are three distinct and identifiable materials. Slag is a molten product of smelting, cinder an infusible or partially fused mass produced at a particular temperature of furnace operation, and bear a material resulting from the reaction of molten smelting products with the refractory lining of the furnace hearth".

According to Wynne and Tylecote (1958), cinders and slags are very similar in their chemical composition (about 60-70% iron oxides and about 20% silica), but they differ greatly in their physical appearance; and it is by this means that they are distinguished. Cinders are very porous, and often the shape of individual ore lumps in them can still be distinguished. In this respect cinders resemble the furnace conglomerate of copper smelting mentioned at the beginning of this chapter. Although some low melting point constituents have bonded the conglomerate particles together, the remains of pieces of charcoal or wood can be seen embedded in the cinder. Cinders are partially smelted ores which have been in such a position in the furnace that reduction has been unable to proceed to completion. The 'genuine' bloomery slags, on the contrary, are black with smooth, rounded surfaces and denser than cinders. They have been formed as other tap slags by the fluxing of gangue materials with iron oxides to give a free-flowing slag of low viscosity.

Though there is no such thing as THE bloomery process, its many variations have something in common as far as slagging is concerned; cf. Osann (1971) and Pelet (1976).

Quoting Tylecote (1962): "As much as possible of the slag has to be removed from iron by liquation, and therefore the smelting process must take place at or above the temperature at which the slags become sufficiently fluid to drain away from the solid iron. This temperature is about 1150°C, well above the minimum temperature at which iron oxide can be reduced to iron, but substantially below the melting point of the metal. Thus, iron is produced in the solid state as a sponge or raw bloom from which the slag partly drains away. The rest is removed by hammering while the slag is still in a fluid state ... After the slag has been squeezed out by gently hammering, forging can be continued at a lower temperature" (under 1050°C). Among the numerous investigators of bloomery slags, the author has drawn valuable information from the following (in addition to those already quoted): Khevenhüller *et al*. (1961), Neumann and Klemm (1949), and Oelsen and Schürmann (1954).

Iron slags may not only come from the actual smelting; iron working (smithing, forging) also produces slags as waste. As is known to everyone working in this field, distinction between the two types is difficult. The presence of ore and/or tap slag at a particular site is evidence of iron smelting. The occasional stray find, particularly in or near settlements, may indicate smithing.

During the early stages of smithing a good deal of slaggy material will be removed from the bloom and such material resembles, chemically and physically, smelting slag and cinders. Later on, the material removed during smithing consists mainly of fairly pure oxidised iron (hammer scale).

Among the examples illustrated in Plates XVI to XXVII are smelting as well as smithing slags.

The chemical analyses of all the samples illustrated in Plates XVI to XXVII are summarised in Table 1.

Plate XVI: The slag found in the ruins of a Roman villa near Berchtesgaden, southern Germany, is a typical tap slag. Its high Mn and low CaO contents point to smelting of manganese-rich ores in a bloomery hearth. The blue-black piece, weighing 250 grams, is dense, rather heavy and partly magnetic.

XVIa) Outer appearance, size 50 x 95 mm;

XVIb) Fayalite crystals in a glass matrix plus wüstite (x200);

XVIc) Same as XVIb), exhibiting beautiful wüstite dendrites (x400).

Plate XVII: This is a very porous bloomery slag from a Roman settlement at Pöcking near Starnberg, southern Germany. Tapping is very unlikely. It is low in iron, high in silica and alumina. The greyish black piece is of low specific gravity, yet partly magnetic (weight 197 grams).

XVIIa) Outer appearance, size 86 x 65 mm;

XVIIb) Fayalite crystals (dark grey) embedding tiny flakes and wires of metallic iron (x150);

XVIIc) Fayalite with large crystals of partly resorbed wüstite and/or magnetite (x150);

XVIId) A silica-rich eutectoid matrix with droplets of metallic iron (x150).

Plate XVIII: Near an early medieval settlement dating to the 7th century AD near Mühltal on the Isar, southern Germany, many isolated pieces of slag were found, but no slag heap. The blue-black specimen is porous, partly magnetic and heavy. The slag, low in silica and high in lime, is probably a wüstite-rich smithing slag (weight 472 grams).

XVIIIa) Outer appearance, size 105 x 92 mm;

XVIIIb) Wüstite in slag matrix (150).

Plate XIX: The iron mining and smelting area at the Ofenpass near Graubünden, Switzerland, was in operation from early (pre-Roman?) times until the 18th century AD. The small piece (weighing 14 grams) is of unknown age, weakly magnetic, rather low in iron oxide but with high contents of metallic iron, lime, alumina and magnesia. The specimen was probably contaminated with furnace lining.

XIXa) Outer appearance, size 29 x 21 mm;

XIXb) Iron inclusions (x75);

XIXc) Fine crystals of fayalite (x150);

XIXd) Fayalite and wüstite 'skeletons' (x150);

XIXe) Metallic iron inclusions showing ferrite with some perlite in coagulated, 'degenerated' form; this is typical for bloomery iron low in Si and Mn (x375).

Plate XX: According to Stursberg (1964), in the valley of the Mul near Hückeswagen, northwest Germany, more than 140 slag heaps, varying in diameter between 15 and 20 metres, have been discovered together with relics of bloomery furnaces. They are all of medieval age. A piece of slag from one of these heaps is illustrated in Plate XX. The findings of the micrographs were verified by X-ray diffraction. The powder pattern indicated the presence of fayalite, quartz and possibly a glass phase.

XXa) Fayalite crystals (x100);

XXb) Same as XXa); centre of sample (x200);

XXc) Same as XXa); rim of sample (x200);

XXd) Same as XXa), plus particles of metallic iron (x500).

Plate XXI: The specimen from Michelsberg near Kelheim in Bavaria, southern Germany, is apparently a partly reduced conglomerate of ore and slag contaminated with furnace lining. Its content of trivalent iron is high, so is its lime and carbonate. The non-magnetic piece with a light brown crust (inside dark) is from a bloomery hearth of unknown age, probably medieval; sample weight 53 grams;

XXIa) Outer appearance, size 38 x 34 mm;

XXIb) Porous structure without any obvious orientation of partly oxidic, partly siliceous composition. Evidently ore fragments mixed with furnace lining (x100);

XXIc) Another section of the same sample (x100).

Plate XXII: The late Hittite site of Şirzi, southeast Turkey (assumed age: 8th century BC), has revealed iron slags as well as ore. The material was published by Bachmann (1967). Specimens from the same site were later investigated by Naumann (unpublished). The results of both investigators are in close agreement. The chemical analyses are summarised in Table 1. The black, heavy, slightly porous, ferro-magnetic slags count among the earliest iron slags investigated so far. Their total iron contents place them in the region of bloomery slags, though their SiO_2 content is surprisingly low.

XXIIa) Outer appearance of one sample, size 50 x 33 mm, weight 62 grams;

XXIIb) Dendrites of wüstite with a little metallic iron in silicate matrix (x50);

XXIIc) Same as XXIIb) (x50);

XXIId) Same as XXIIb) (x100);

XXIIe) Same as XXIIb) (x100);

XXIIf) Same as XXIIb) (x200);

XXIIg) Same as XXIIb) (x200).

Plate XXIII: From the late Hittite site at Şirzi, southeast Turkey (cf. Plate XXII), a piece of rich hematite ore with some calcite as gangue is included in this catalogue, because we have here one of the rare instances where slag and ore are found together, so we know exactly the source of iron ore from which the slags were derived.

XXIIIa) Outer appearance of sample, size 55 x 24 mm, weight 21 grams;

XXIIIb) Hematite particles contaminated with gangue (calcite) (x50);

XXIIIc) Same as XXIIIb) (x100);

XXIIId) Same as XXIIIb) (x200).

Plate XXIV: This is a piece of bloomery slag. The dense, heavy, blue-black specimen with signs of magnetism is a stray find near the circular wall of Michelsberg near Kelheim, southern Germany, of unknown age (Celtic?). It shows a remarkably high manganese content.

XXIVa) Outer appearance of sample, size 74 x 55 mm, weight 240 grams;

XXIVb) Clearly visible are two eutectics in the system FeO/SiO_2 composed of wüstite and fayalite (x100);

XXIVc) Same as XXIVb) (x100).

Plate XXV: A piece of partly reduced Mn-rich 'Bohnerz' (ie. botryoidal limonite, $Fe_2O_3.nH_2O$) from the same site as the slag illustrated in the previous plate is another example of ore and slag from one site. The rounded ore nodule is of dark brown colour, of low specific gravity and non-magnetic.

XXVa) Outer appearance of sample, size 29 x 24 mm, weight 16 grams;

XXVb) Centre part of reduced 'Bohnerz' particle (x100);

XXVc) Rim section of same specimen (x100).

Plate XXVI: Stray finds from Eggersberg Castle near Kelheim, southern Germany, include four specimens that were analysed chemically (*cf.* Table 1). Only sample No. 3 is illustrated here by a set of micrographs. The set of samples was characterised by Naumann (unpublished) as follows:

No. 1: Blue-black, kidney shaped, porous tapping slag, weakly magnetic, weight 199 grams, low in iron, high in manganese, SiO_2 and Al_2O_3 above average, low in lime. Under the microscope primary crystallised fayalite is distinguishable, together with a fayalite-silica eutectoid and some metallic iron (not illustrated).

No. 2: Porous furnace (?) slag, brown, porous, low specific gravity, in some areas noticeably magnetic, weight 27 grams, low in FeO, high in MnO, SiO_2 and Al_2O_3, very high in BaO and TiO_2. Microscopic characterisation same as No. 1 (not illustrated).

No. 3: (ie. the specimen illustrated in Plate XXVI) black tapping slag, heavy, varying degree of magnetism, weight 103 grams, high in MnO and Al_2O_3.

No. 4: Deep black furnace slag with micropores, partly magnetic, weight 30 grams, high in MnO and Al_2O_3, very similar to No. 3 (not illustrated).

XXVIa) Outer appearance of sample No. 3, size 49 x 40 mm;

XXVIb) Fayalite and wüstite, plus two eutectoids (x100);

XXVIc) Same as XXVIb), different area of the section (x100).

Plate XXVII: The last plate of the catalogue is entirely dedicated to a series of smithing slags from a Mamluk smithy (11th to 12th century AD) near Eski Kâhta, southeast Turkey, plus a recent smithing slag from the same site included for comparison. A full description of the material has been given by Bachmann (1970).

XXVIIa) Smithing slag with metallic iron and trivalent iron. α-Fe (white) in a grey partly decomposed (iron hydroxides) matrix (x50);

XXVIIb) Same as XXVIIa) (x50);

XXVIIc) Porous, sponge iron in a grey, layered oxidic matrix (x50);

XXVIId) Slag with predominantly metallic iron (ie. low in iron oxides). Droplets of iron are dispersed in a silicate matrix partly glassy and partly crystallised (x50);

XXVIIe) Another slag rich in metallic iron with additional oxide and silicate phases (oxides: wüstite and goethite, formed by weathering; silicates probably present as calcium-iron-silicates like hedenbergite, $CaFeSi_2O_6$ and/or kirschsteinite, $CaFeSiO_4$ (x50);

XXVIIf) Modern smithing slag from the smithy in the Kurdish village of Eski Kâhta with metallic and divalent iron, etched with 'nital', with calcium iron compounds like $Ca_2Fe_2O_5$ (x50);

XXVIIg) Same as XXVIIf) (x100);

XXVIIh) Same as XXVIIf) (x100).

Acknowledgements

Thanks to the initiative and endeavour of Professor B. Rothenberg, the Institute for Archaeo-metallurgical Studies (IAMS) was founded and subsequently made a body affiliated to the Institute of Archaeology of the University of London. The author is most obliged to Professor Rothenberg for the invitation to participate in the lecturing programme set up by IAMS; otherwise this booklet would and could not have been written.

For accepting this monograph in the publication programme of the Institute of Archaeology, University of London, I owe sincere thanks to its Director, Professor J. D. Evans. Throughout the preparation of the manuscript, I enjoyed continuous encouragement and assistance from Dr. N. J. Seeley, Head of the Department of Conservation and Materials Science of the Institute of Archaeology. He not only checked the draft for wording and style but also made important comments and corrections. Professor R. F. Tylecote provided not only essential information through his outstanding publications but also through personal advice.

The numerous plates come from two sources: A large set was handed to me by the late C. Böhne, made for him by Dr. F. K. Naumann at the Max-Planck-Institut für Eisenforschung, Düsseldorf. I have to thank these people for permission to make free use of their material. The other source is the author's archive, to which a number of fellow workers in various departments of Degussa (Deutche Gold- und Silberscheideanstalt, Frankfurt/Main) have contributed. Thanks are due to these colleagues and to the company's management.

Recommended reading and references

Slag analyses and investigations are scattered widely through the archaeological, archaeometallurgical and technical literature. The vast number of journals in which such research is likely to appear necessitates the consultation of an abstract service. The *Journal of the Historical Metallurgical Society*, affiliated to the Metals Society, now in its 15th volume, is not only dedicated to the publication of articles on ancient and historical metallurgy, but supplies world wide abstracts on this subject as well.

Among the references listed are included a number of books and articles which are not cited in the text. They provide basic information on archaeometallurgy in general and in some of them analytical data on slags are quoted. These recommended publications are marked with an asterisk. To some readers this selection may seem biased, but even if the selection is subjective, cross-references lead to other sources of information.

Allchin, F. R. 1963 *Neolithic cattle-keepers of South India: a study of the Deccan ash mounds*. Cambridge: Cambridge University Press.

Bachmann, H. G. 1967 Untersuchung einer vermutlich hethitischen Eisenschlacke aus Şirzi (Malatya), Türkei. *Archiv f. d. Eisenhüttenwesen* 38: 809-812.

*Bachmann, H. G. 1968 Untersuchungen an vor- und frühgeschichtlichen Kupferverhüttungsschlacken. *Erzmetall* 21: 419-424.

Bachmann, H. G. 1970 Eisenschlacken aus Eski Kâhta (Vil. Adiyaman), Südosttürkei. *Archiv f. d. Eisenhüttenwesen* 41: 731-736.

Bachmann, H. G. 1974 Untersuchungen an frühgeschichtlichen Kupferverhüttungsschlacken aus Zypern. *Leitz-Mitteilungen f. Wissenschaft u. Technik*, Suppl. Band I: 177-180.

Bachmann, H. G. 1976 Copper slags; preliminary report. *Hala Sultan Tekke 1; Excavation, 1897-1971* (ed. P. Åström, D. M. Bailey and Vassos Karageorghis). Göteborg: Studies in Mediterranean Archaeology, vol. XLV. 127-128.

Bachmann, H. G. 1978 The phase composition of slags from Timna site 39. *Chalcolithic Copper Smelting*. London: IAMS Monograph No. 1. 21-23.

*Bachmann, H. G. 1980 Early copper smelting techniques in Sinai and Negev as deduced from slag investigations. *Scientific Studies in Early Mining and Extractive Metallurgy* (ed. P. Craddock). London: British Museum Occasional Paper No. 20. 103-134.

Bachmann, H. G. and Rothenberg, B. 1980 Die Verhüttungsverfahren von Site 30. *Antikes Kupfer im Timna-Tal in der Arabah (Israel)* (ed. H. G. Conrad and B. Rothenberg). Bochum: Der Anschnitt, Beiheft 1. 215-236.

Biek, L. 1970 'Vitrified products' - a comment. *Bull. Hist. Met. Group* 4: 81-82.

Böhne, C. 1968 Über die Kupferverhüttung der Bronzezeit, Schmelzversuche mit Kupferkieserzen. *Archaeologica Austriaca* 44: 49-60.

Coghlan, H. H. 1956 *Notes on prehistoric and early iron in the Old World.* Oxford: Pitt Rivers Museum Occasional Papers on Technology, 8.

*Czedik-Eysenberg, F. 1958 Beitrage zur Metallurgie des Kupfers in der Urzeit. *Archaeologica Austriaca* 3: 1-18.

Dana, J. D. 1951 *System of Mineralogy*, Vol. II. New York: J. Wiley. 62-69.

Evans, R. T. and Tylecote, R. F. 1967 Some vitrified products of non-metallurgical significance. *Bull. Hist. Met. Group* 1: 22-23.

*Faber, W. 1954 Mikroskopie der Metallhüttenschlacken. *Handbuch der Mikroskopie in der Technik*, Band II, Teil 2 (ed. H. Freund). Frankfurt am Main: Umschau Verlag. 519-594.

*Forbes, R. J. 1971 *Studies in Ancient Technology*, Vol. VIII. Leiden: E. J. Brill.

*Forbes, R. J. 1972 *Studies in Ancient Technology*, Vol. IX. Leiden: E. J. Brill.

Hampl, F. 1953 Kupfererzbergbau im südöstlichen Niederösterreich. *Archaeologica Austriaca* 13: 47-72.

*Healy, J. F. 1978 *Mining and Metallurgy in the Greek and Roman World.* London: Thames and Hudson.

Hetherington, R. 1980 Investigations into primitive lead smelting and its products. *Aspects in Early Metallurgy* (ed. A. Oddy). London: British Museum Occasional Paper No. 17. 27-40.

Khevenhüller, G., Mitsche, R. and Trojer, F. 1961 Aufbau dreier römischer Eisenluppen und der damit verbundenen Schlackenbestandteile vom Magdalensberg, Kärnten. *Carinthia* II, 71: 81-87.

Laub, G. 1970/71 Zur Frage eines Altbergbaues auf Kupfererze im Harzgebiet. *Harz-Zeitschrift* 22/23: 97-143.

Laub, G. 1975 Der Bleifund vom Radauberg bei Bad Harzburg. *Harz-Zeitschrift* 27: 31-56.

Laub, G. 1978 Bemerkungen zu den Unterscheidungsmerkmalen alter Schlacken aus der Verhüttung von Rammelsberger Erzen. *Harz-Zeitschrift* 30: 107-112.

*Levin, E. M., Robbins, C. R. and McMurdie, H. F. 1974 *Phase Diagrams for Ceramists* (ed. M. K. Reser). 3rd ed. Columbus, Ohio: The American Ceramist Society. 5-29.

Lupu, A. 1970 Metallurgical Aspects of Chalcolithic Copper Working at Timna (Israel). *Bull. Hist. Met. Group* 4: 21-23.

*Maréchal, J. R. 1962 *Considérations sur la métallurgie préhistorique.* Lammersdorf n. Aachen.

Mayrhofen, R. 1953 Geologische, mineralogische und technologische Beobachtungen und Untersuchungen auf den urzeitlichen Preiner Schmelzplätzen und ihrem Assoziationskreis. *Archaeologica Austriaca* 13: 73-104.

Miles, H. 1975 Barrows on the St. Austell Granite, Cornwall. *Cornish Archaeology* 14: 5-81.

*Milton, C. Dwornick, E. J., Finkelman, R. B. and Toulmin III, P. 1976 Slag from an ancient copper smelter at Timna, Israel. *J. Hist. Metall. Soc.* 10: 24-33.

Morton, G. R. and Wingrove, J. 1969a Slag, cinder and bear. *Bull. Hist. Met. Group* 3: 55-61.

*Morton, G. R. and Wingrove, J. 1969b Constitution of Bloomery Slags: Part I. Roman. *J. Iron and Steel Inst.* 207: 1556-1564.

*Morton, G. R. and Wingrove, J. 1972 Constitution of Bloomery Slags. Part II: Medieval. *J. Iron and Steel Inst.* 210: 478-488.

Needham, J. 1958 *The Development of Iron and Steel Metallurgy in China.* Cambridge: Heffer (for the Newcomen Society).

Neumann, B. and Klemm, H. 1949 Der Gefügeaufbau der Eisenoxyd-Silikate der Rennfeuer- u. Stückofenschlacken und die Vorgänge bei der Ausscheidung des metallischen Eisens aus solchen Schlacken. *Archiv f. Metallkunde*, Heft 1: 7-11.

*Obenauer, K. 1954 Mikroskopische Untersuchungstechnik der Eisenhüttenschlacken. *Handbuch der Mikroskopie in der Technik*, Band II, Teil 2, (ed. H. Freund). Frankfurt am Main: Umschau-Verlag. 459-518.

*Oelsen, W. and Schürmann, E. 1954 Untersuchungsergebnisse alter Rennfeuerschlacken. *Arch. Eisenhüttenwesen* 25: 507-514.

*Osann, B. 1971 *Rennverfahren und Anfänge der Roheisenerzeugung.* Düsseldorf: Verein Deutscher Eisenhüttenleute.

*Otto, H. and Witter, W. 1952 *Handbuch der altesten vorgeschichtlichen Metallurgie in Mitteleuropa.* Leipzig: J. Ambrosius Barth-Verlag. 28-29.

*Pelet, P. L. 1976 Versuch einer Klassifizierung frühgeschichtlicher Eisenschmelzöfen. *Arch. Eisenhüttenwesen* 47: 709-712.

Presslinger, H., Eibner, C., Walach, C. and Sperl, G. 1980 Ergebnis der Erforschung urnenfelderzeitlicher Kupfermetallurgie im Paltental. *Berg- u. Hüttenmännische Monatshefte* 125: 131-142.

Rosenqvist, T. 1974 *Principles of Extractive Metallurgy*. McGraw Hill Kogusha Ltd., Int. Student Edition.

Rostoker, W. and Sadowski, M. 1980 The carbon reduction of fully oxidised chalcopyrite (copper) ores. *J. Hist. Metall. Soc.* 14: 39-42.

*Rothenberg, B. 1972 *Timna, Valley of the Biblical Copper Mines*. London: Thames and Hudson.

Schmiedl, J., Campa, S., Baduar, L. and Tomašek, K. 1975 Physikalische Eigenschaften von Schlackensystemen in der Kupfermetallurgie. *Neue Hütte* 20: 149-152.

*Sperl, G. 1980a *Über die Typologie urzeitlicher, frühgeschichtlicher und mittelalterlicher Eisenhüttenschlacken*. Studien zur Industrie-Archäologie VII. Wien: Österreiche Akademie der Wissenshcaften.

Sperl, G. 1980b Metallographic examination of Bronze Age copper. *Metals Technology*: 212-217.

Steinberg, A. and Koucky, F. 1974 Preliminary metallurgical research in the ancient Cypriot copper industry. *American Expedition to Idalion, Cyprus* (eds. L. E. Stager, A. Walker and G. E. Wright). Cambridge, Mass.: American Schools of Oriental Research. 149-178.

Stursberg, E. E. 1964 *Geschichte des Hütten- und Hammerwesens im ehemaligen Herzogtum Berg*. Beitrage z. Geschichte Remscheids, Heft 8. Remscheid.

*Tite, M. S. 1972 *Methods of Physical Examination in Archaeology*. London and New York: Seminar Press.

*Trojer, F. 1963 *Die oxydischen Kristallphasen der anorganischen Industrieprodukte*. Stuttgart: E. Schweizerbartsche Verlagsbuchhandlung.

*Trojer, F. 1974 Die Mikroskopie der Eisenhüttenschlacken. *Handbuch der Mikroskopie in der Technik*, Band IV, Teil 2. Frankfurt am Main: Umschau-Verlag. 393-437.

*Tylecote, R. F. 1962 *Metallurgy in Archaeology. A Prehistory of Metallurgy in the British Isles*. London: Edward Arnold.

Tylecote, R. F. 1966 The history of the tin industry in Cornwall. *Cornish Archaeology* 5: 30-33.

Tylecote, R. F. 1974 Can copper be smelted in a crucible? *J. Hist. Metall. Soc.* 8: 54.

*Tylecote, R. F. 1976 *A History of Metallurgy*. London: The Metals Society.

Tylecote, R. F. 1979 Unpublished lecture notes.

Tylecote, R. F. 1980 Calenick: A Cornish tin smelter, 1702-1891. *J. Hist. Metall. Soc.* 14: 1-16.

Tylecote, R. F. and Boydell, D. J. 1978 Experiments on copper smelting based on early furnaces found at Timna. *Chalcolithic Copper Smelting, Excavations and Experiments*. London: IAMS Monograph No. 1. 27-49.

Verein Deutscher Eisenhüttenleute 1981 *Schlackenatlas/Slag Atlas* (prepared by the Committee for Fundamental Metallurgy; English and German). Düsseldorf: Verlag Stahleisen m.b.H.

Wagner, G. A. 1981 Personal communication.

*White, J. R. 1980 Historic blast furnace slags: archaeological and metallurgical analysis. *J. Hist. Metall. Soc.* 14: 55-64.

*Wynne, E. J. and Tylecote, R. F. 1958 An experimental investigation into primitive iron smelting technique. *J. Iron and Steel Inst.*: 339-348.

Yazawa, A. 1980 Distribution of various elements between copper, matte and slag. *Erzmetall* 33: 377-382.

Youngblood, E., Frederiksson, B. J., Kraut, F. and Frederiksson, K. 1978 Celtic vitrified forts: implications of a chemical-petrological study of glasses and source rocks. *J. Arch. Sci.* 5: 99-121.

Zeuner, F. E. 1959 On the origin of the cinder mounds of the Bellary district, India. *Bull. Inst. Archaeol. London* 2: 37-44.

Zschocke, K. and Preuschen, E. 1932 *Das urzeitliche Bergbaugebiet von Mühlbach-Bischofshofen*. Materialen zur Urgeschichte Österreichs, 6. Heft. Wien: Selbstverlag der Anthropologischen Gesellschaft. 99-100.

Plate Ia (cf. p.20)

Plate Ib (cf. p.20)
(Each square on the scale = 1 centimetre)

Plate IIa (cf. p.21)

Plate IIb (cf. p.21)
(Each square on the scale = 1 centimetre)

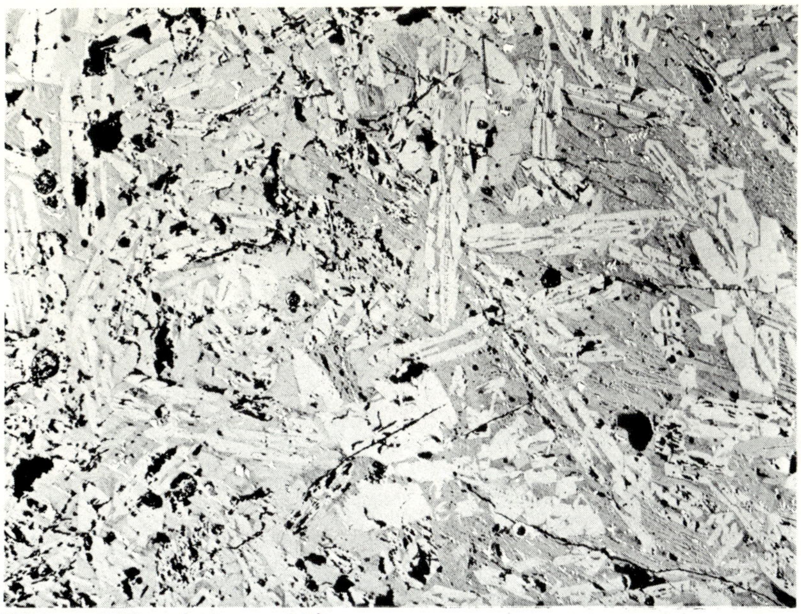
Plate IIIa (cf. p.21)

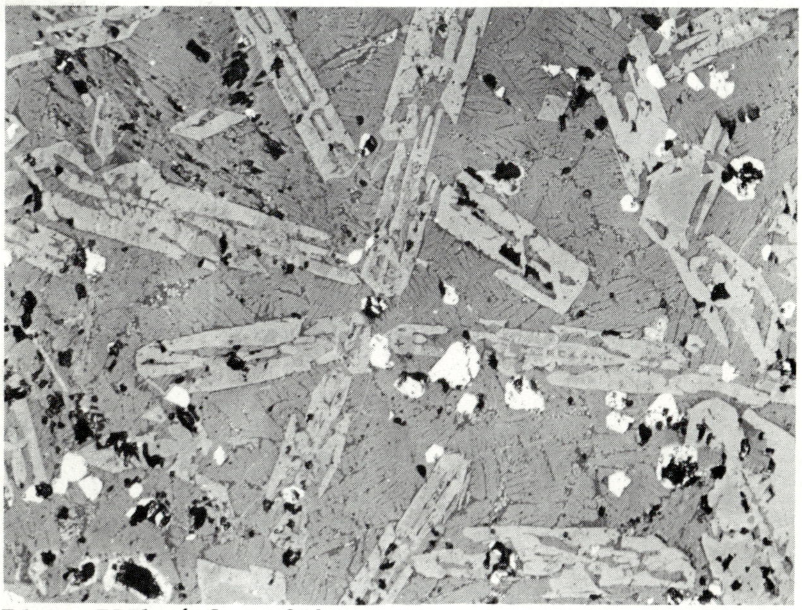
Plate IIIb (cf. p.21)

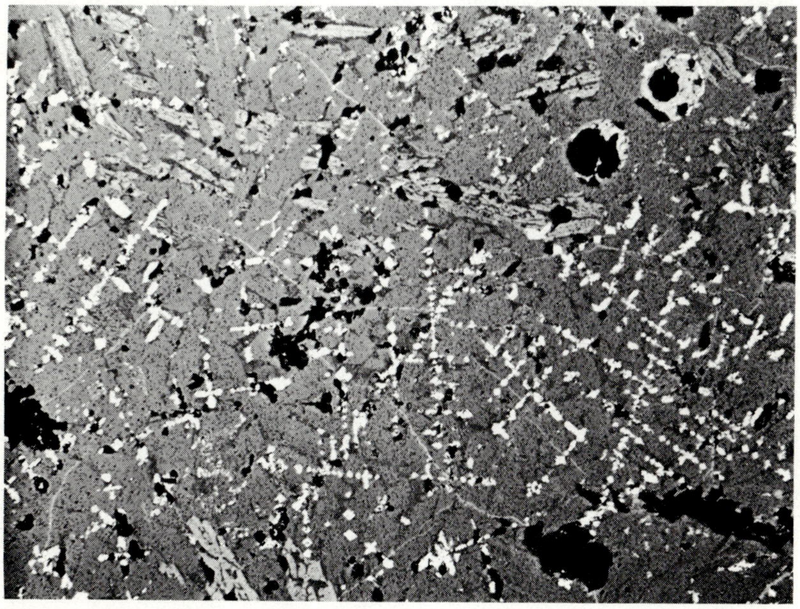
Plate IIIc (cf. p.21-22)

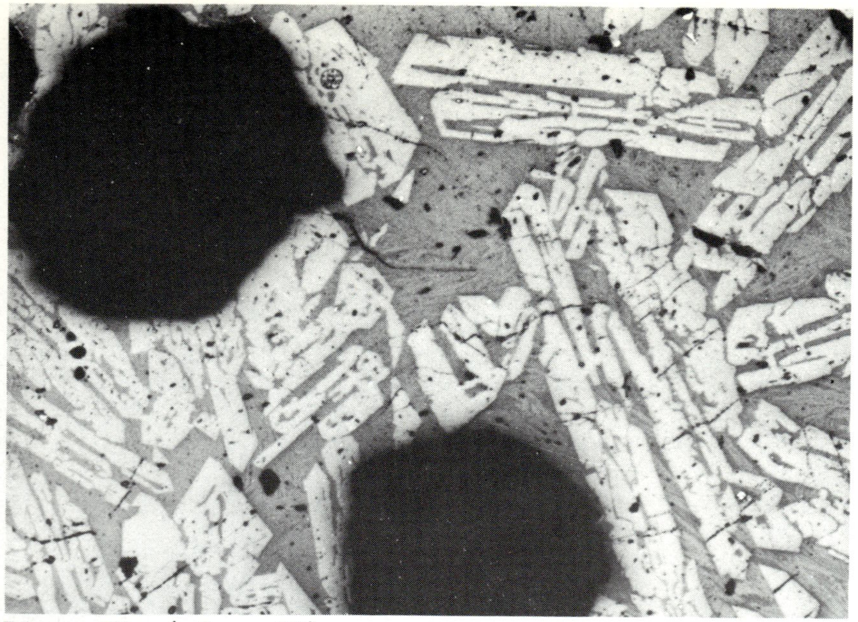
Plate IVa (cf. p.22)

Plate IVb (cf. p.22)

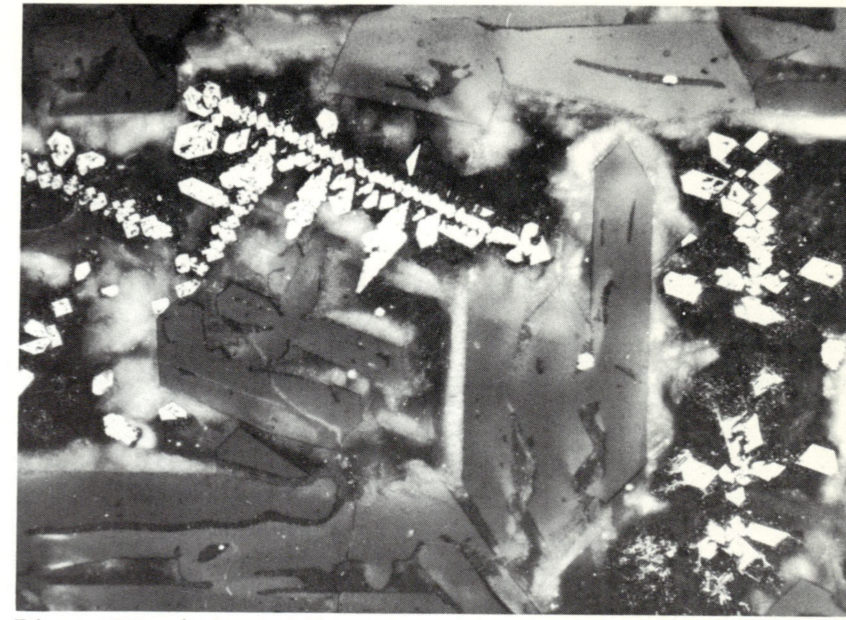
Plate IVc (cf. p.22)

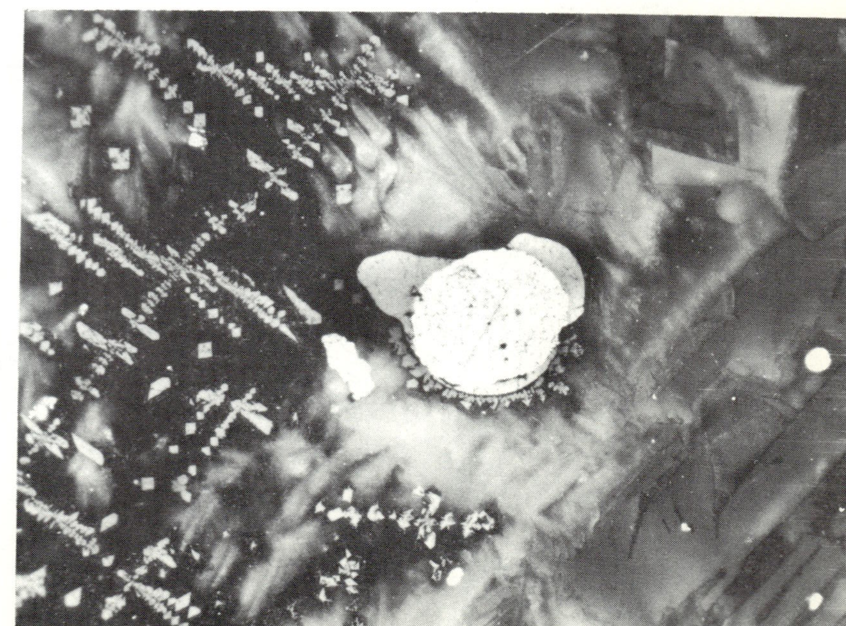
Plate IVd (cf. p.22)

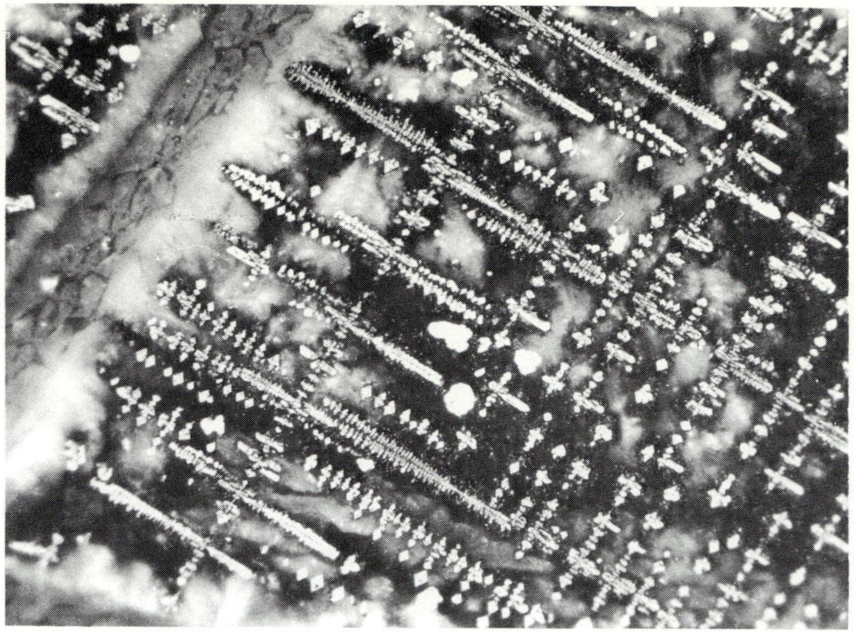

Plate IVe (cf. p.22)

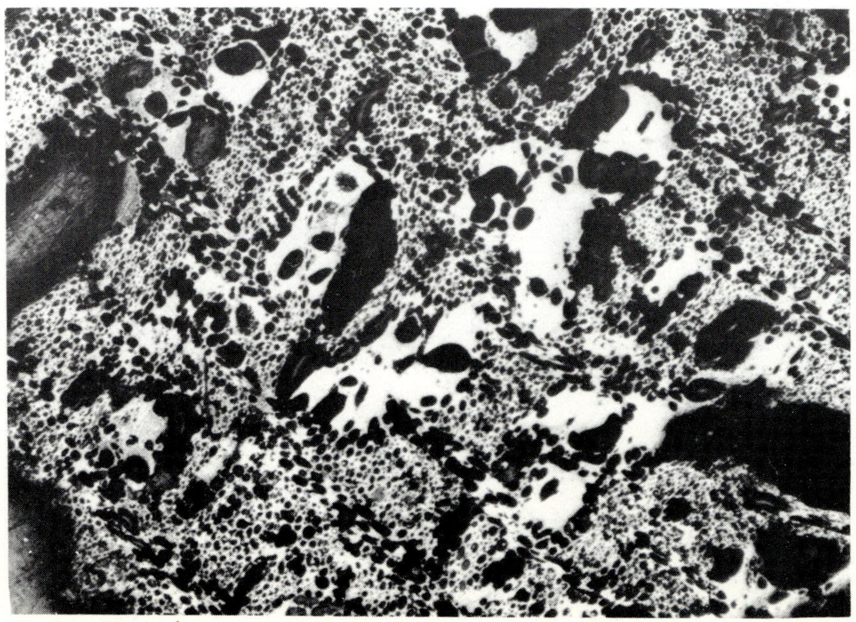

Plate IVg (cf. p.22)

Plate IVf (cf. p.22)

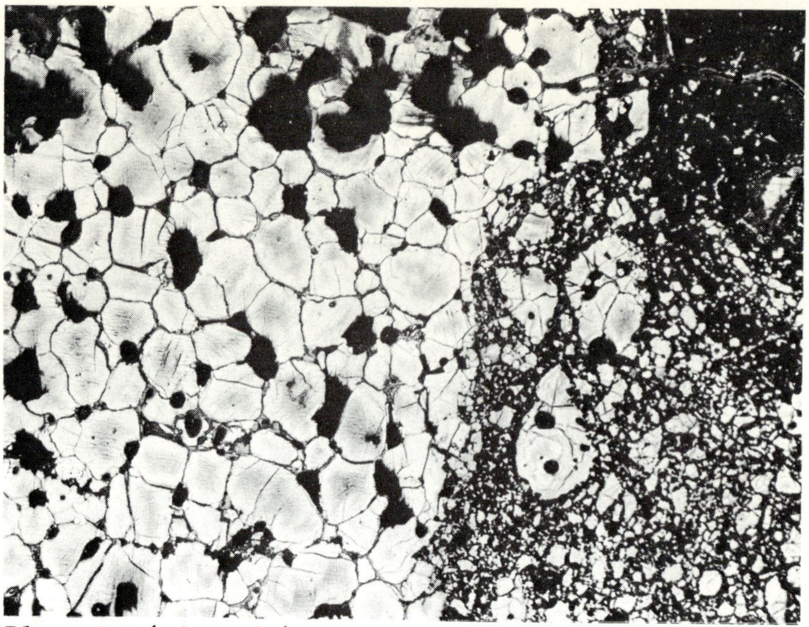

Plate Va (cf. p.23)

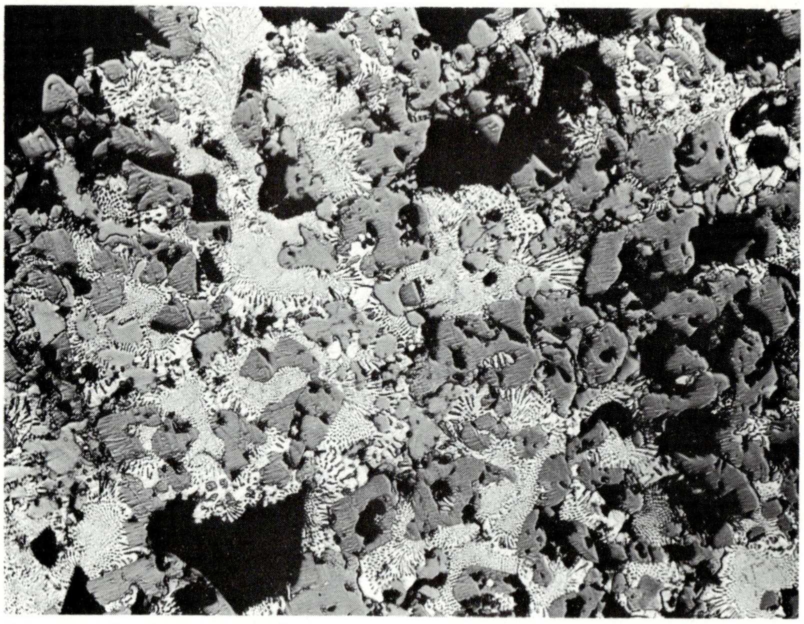

Plate Vb (cf. p.23)

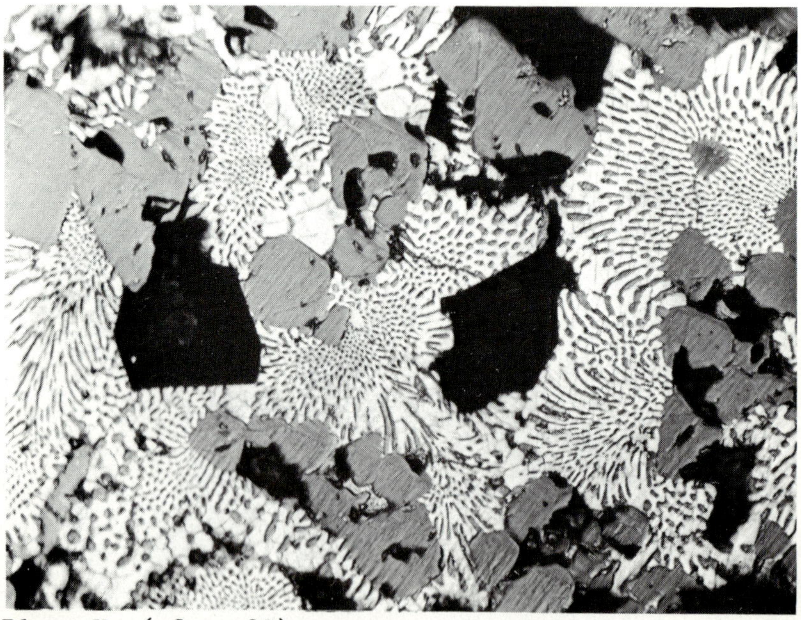

Plate Vc (cf. p.23)

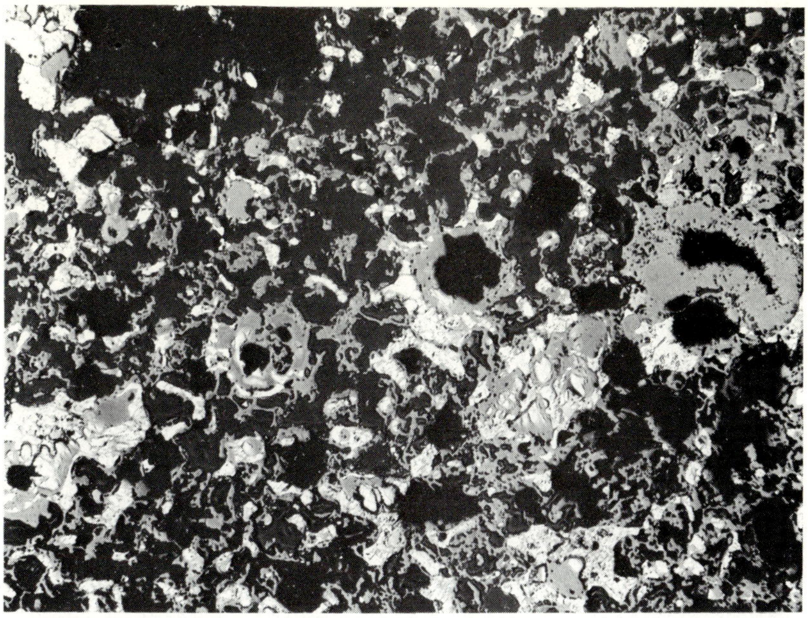
Plate VIa (cf. p.23)

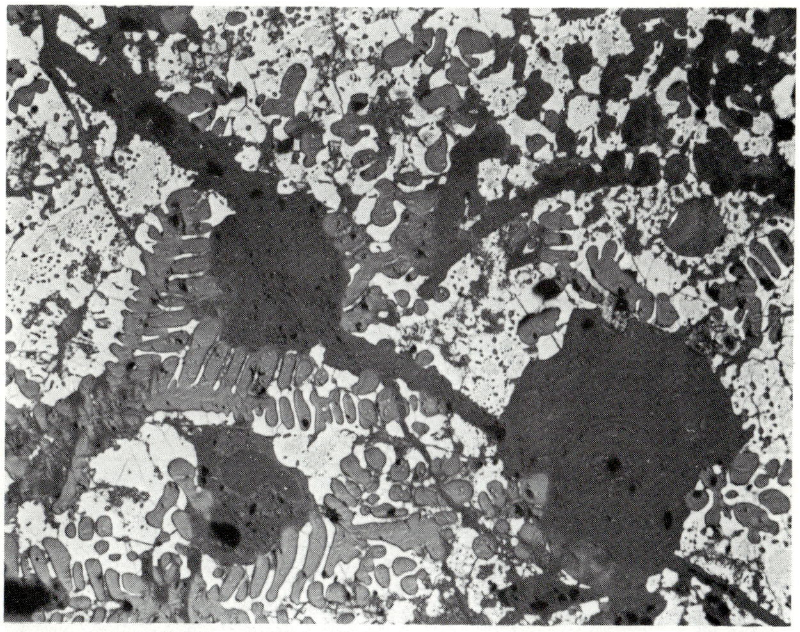
Plate VIb (cf. p.23)

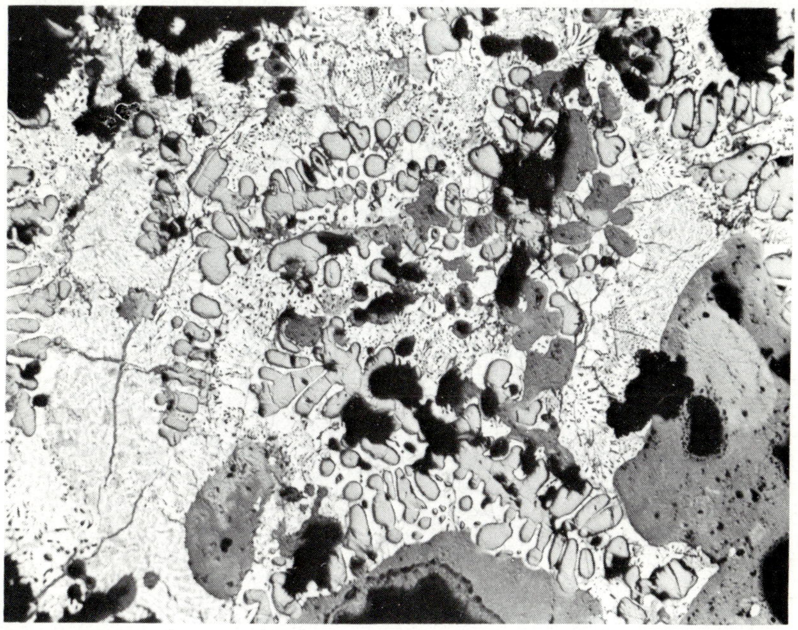
Plate VIc (cf. p.23)

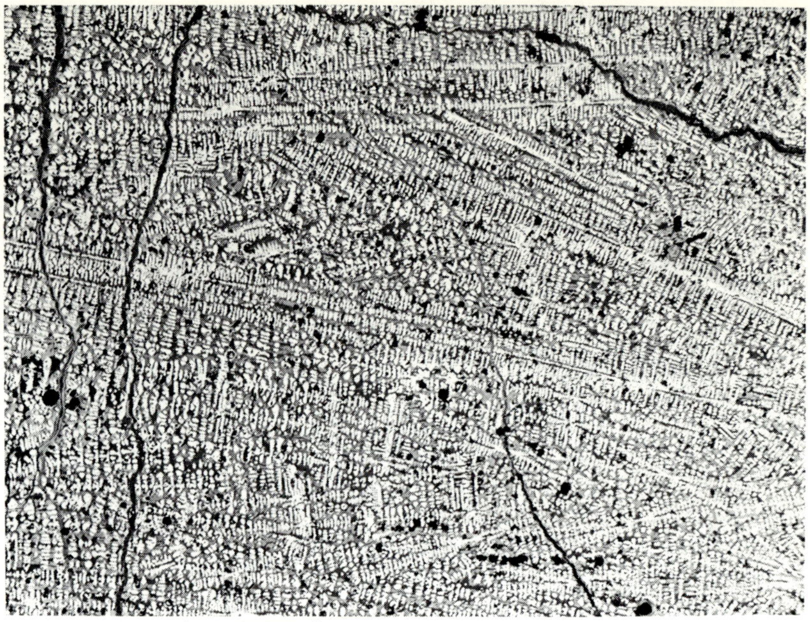

Plate VIIa (cf. p.23)

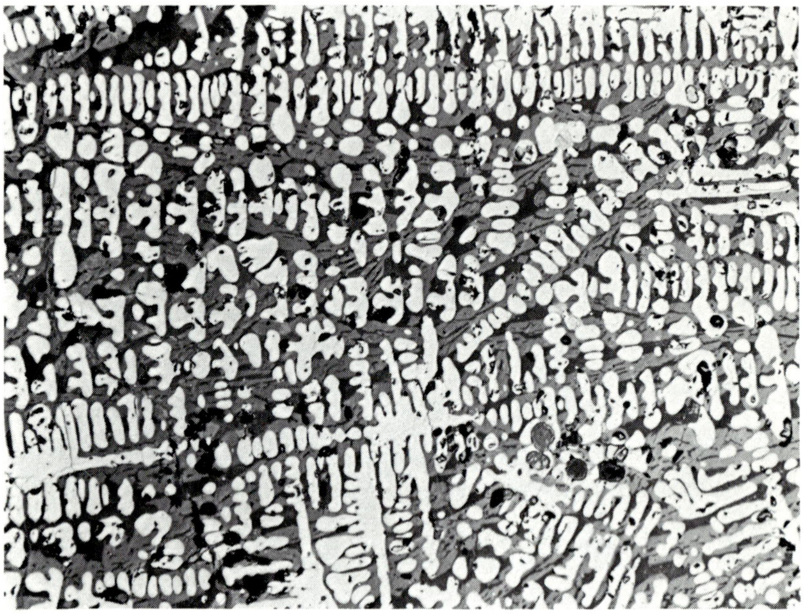

Plate VIIb (cf. p.24)

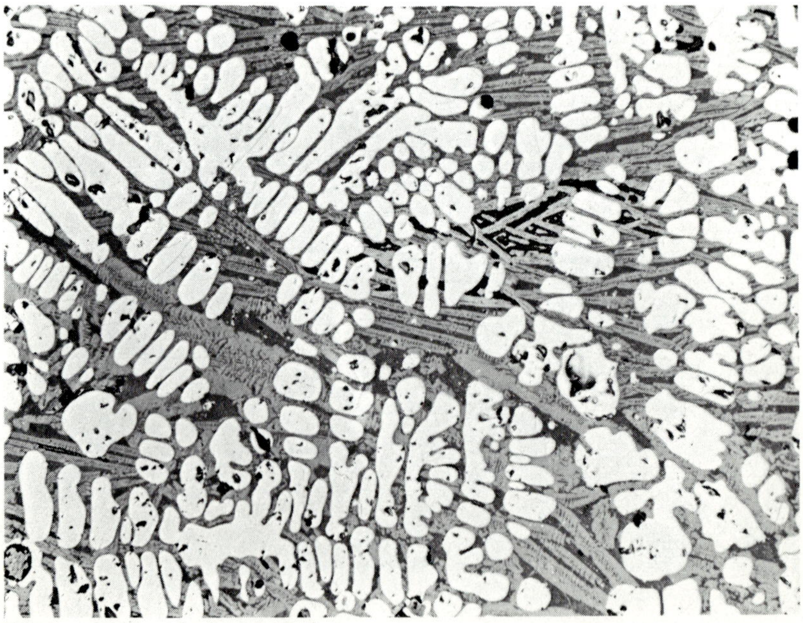

Plate VIIc (cf. p.24)

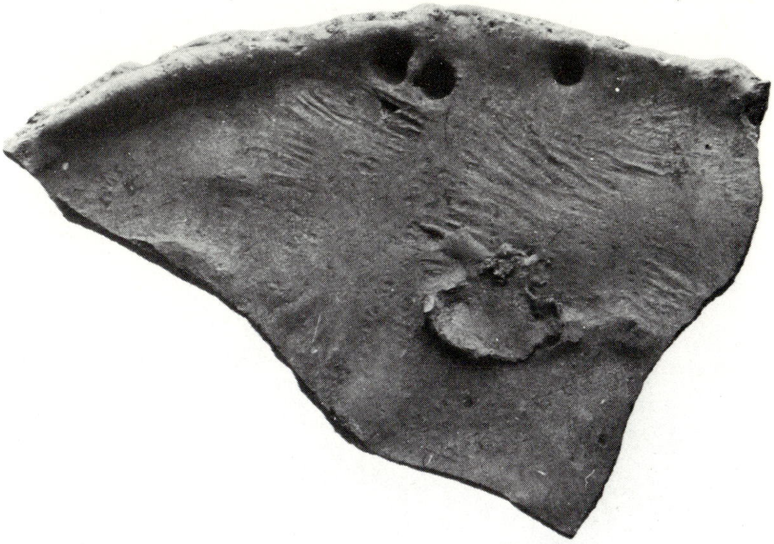

Plate VIIIa (cf. p.24)

Plate VIIIb (cf. p.24)

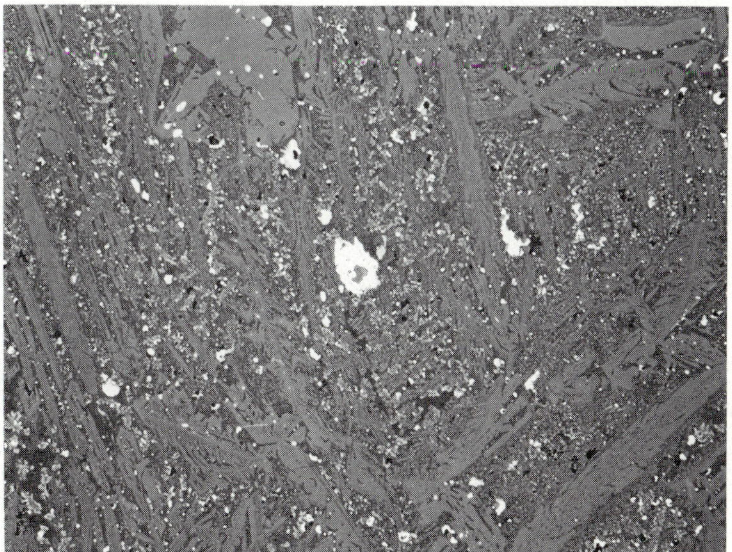

Plate VIIIc (cf. p.24)

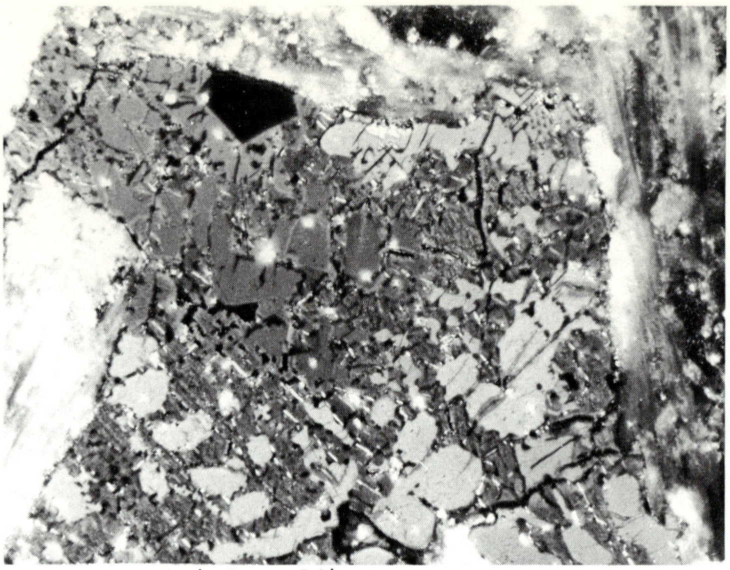

Plate VIIId (cf. p.24)

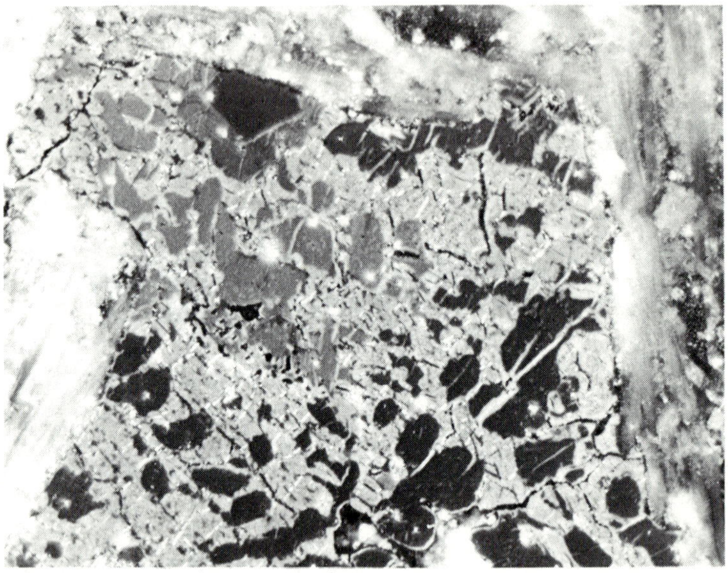

Plate VIIIe (cf. p.23)

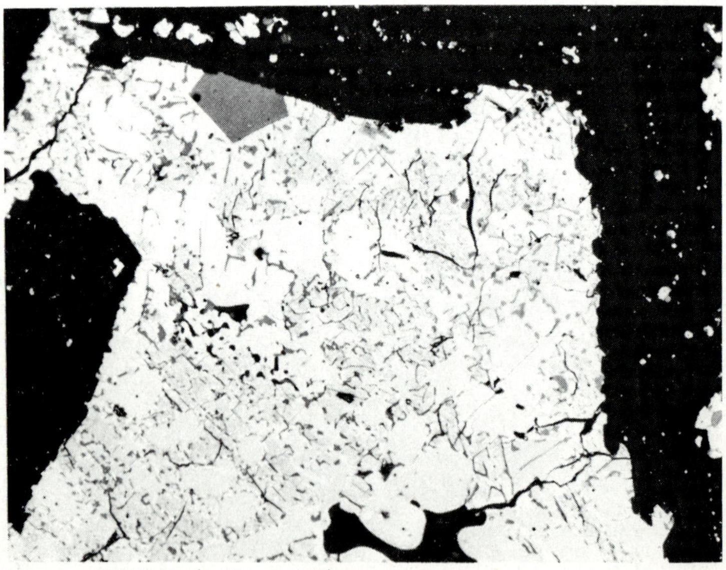

Plate VIIIf (cf. p.24)

Plate IXa (cf. p.24)

Plate IXb (cf. p.24)

Plate IXc (cf. p.24)

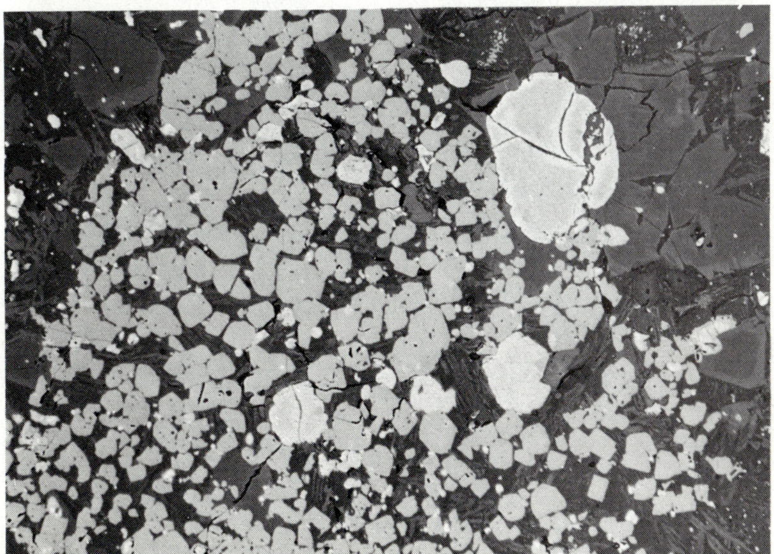

Plate IXd (cf. p.24)

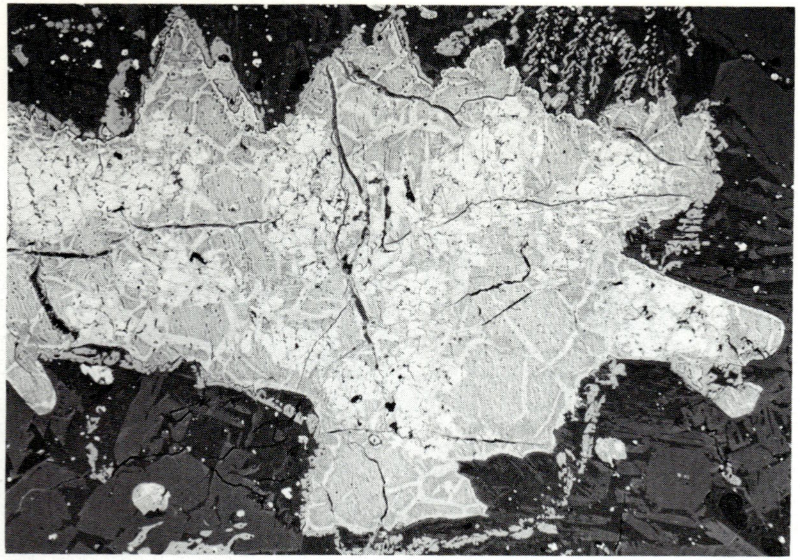

Plate IXe (cf. p.24)

Plate IXe (cf. p.24)

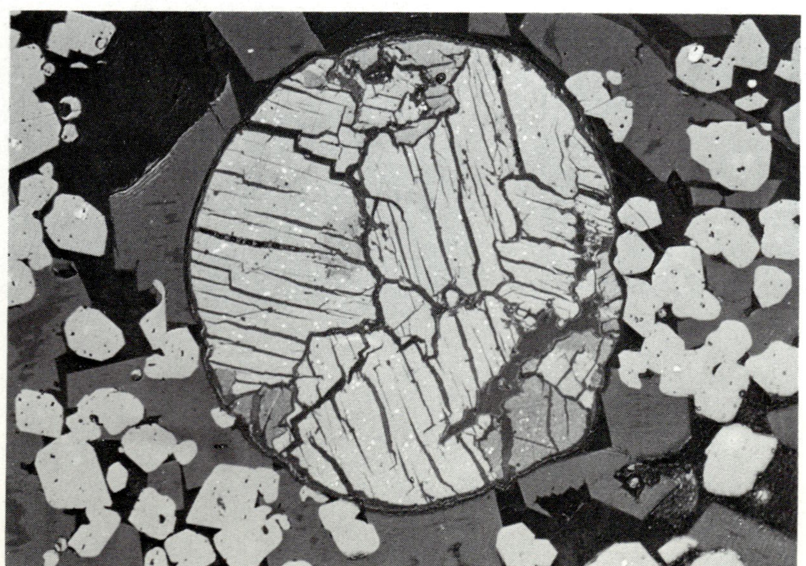

Plate IXf (cf. p.24)

Plate IXf (cf. p.24)

Plate Xa (cf. p.24)

Plate Xb (cf. p.24)

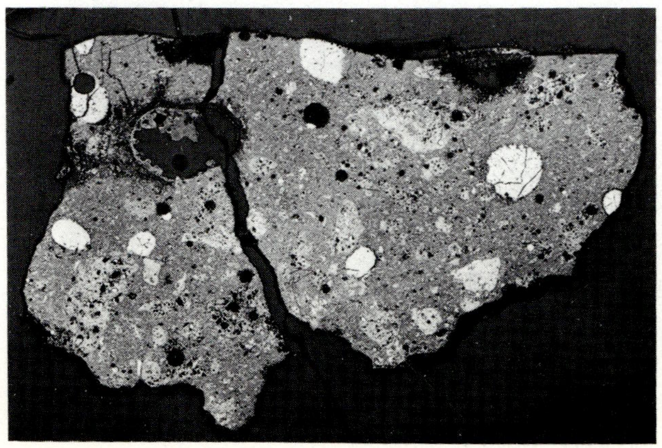

Plate Xc (cf. p.24)

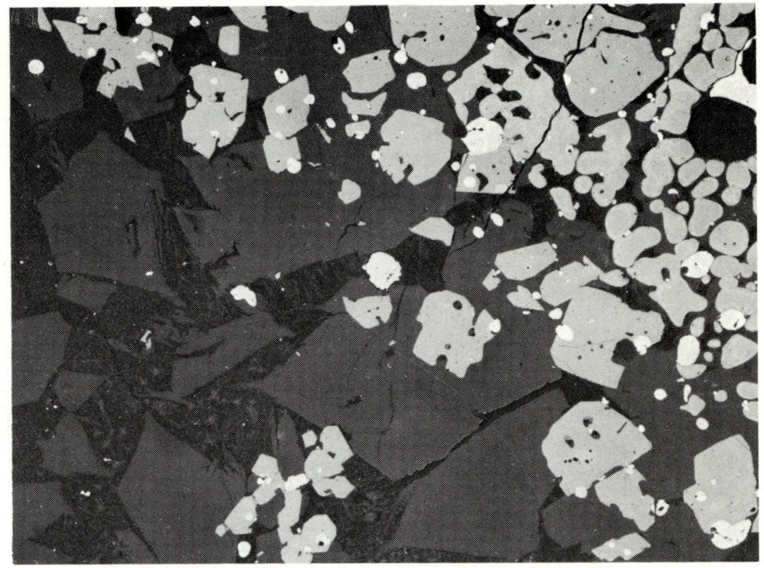

Plate Xd (cf. p.24)

Plate Xg (cf. p.25)

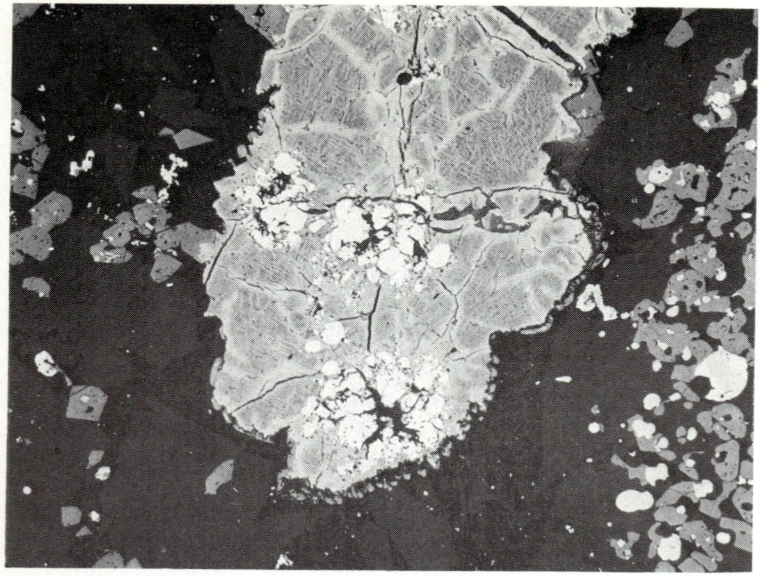

Plate Xe (cf. p.24-25)

Plate Xe (cf. p.24-25)

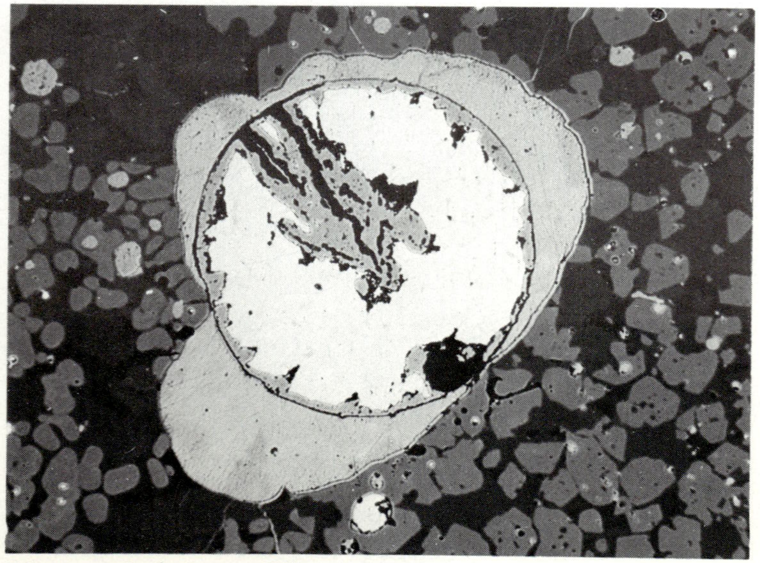

Plate Xf (cf. p.25)

Plate Xf (cf. p.25)

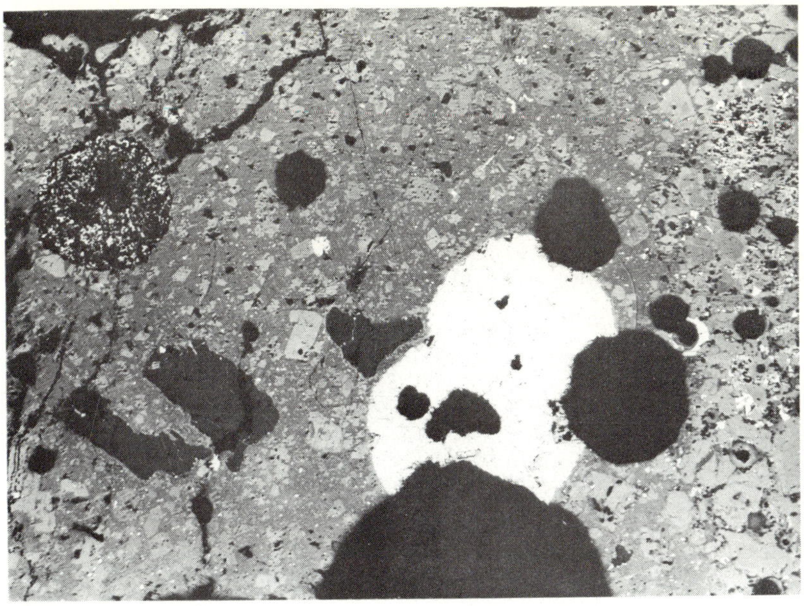

Plate XIa (cf. p.25)

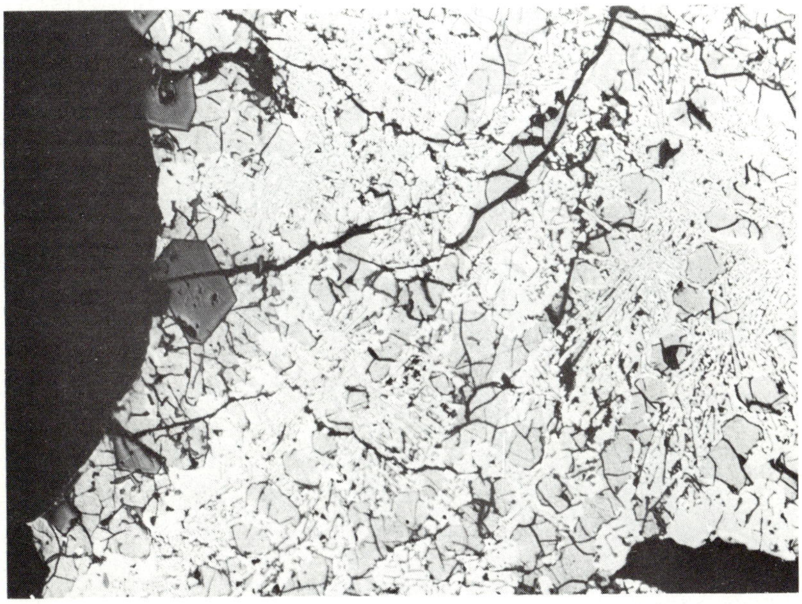

Plate XIb (cf. p.25)

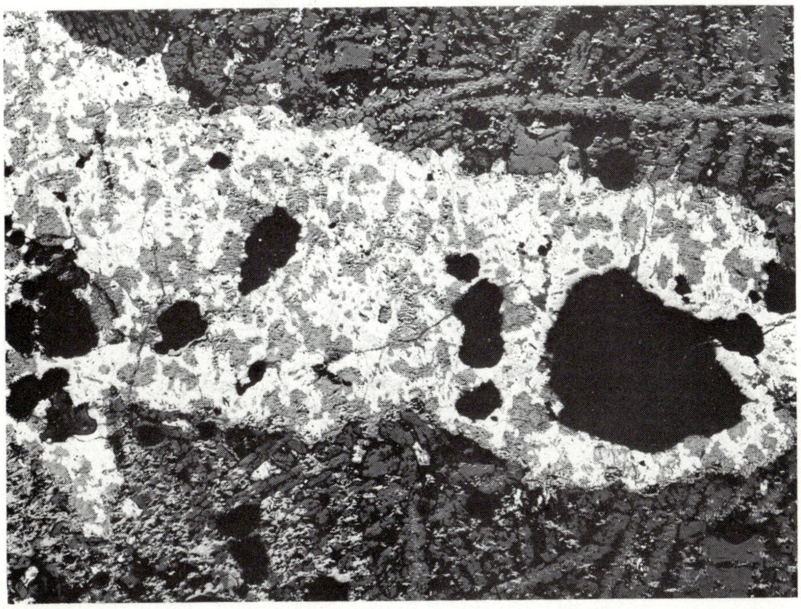

Plate XIc (cf. p.25)

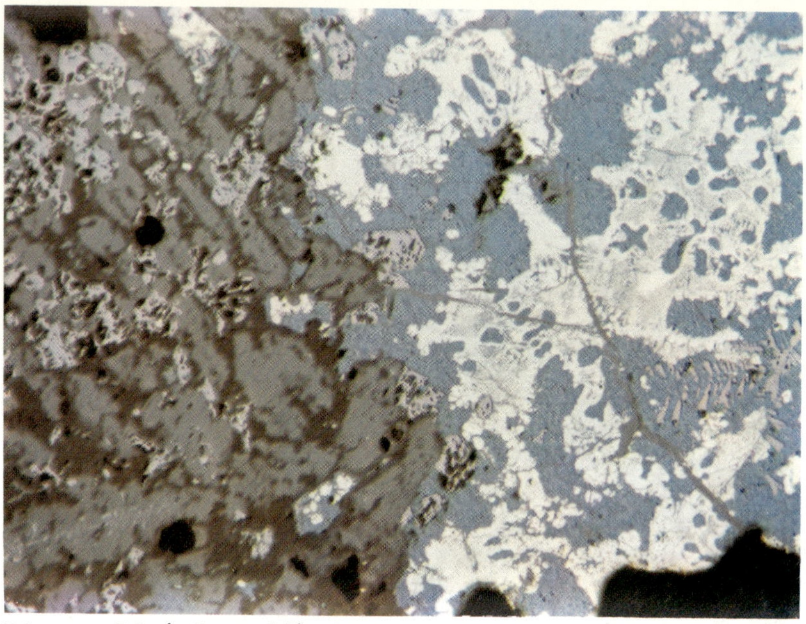

Plate XId (cf. p.25)

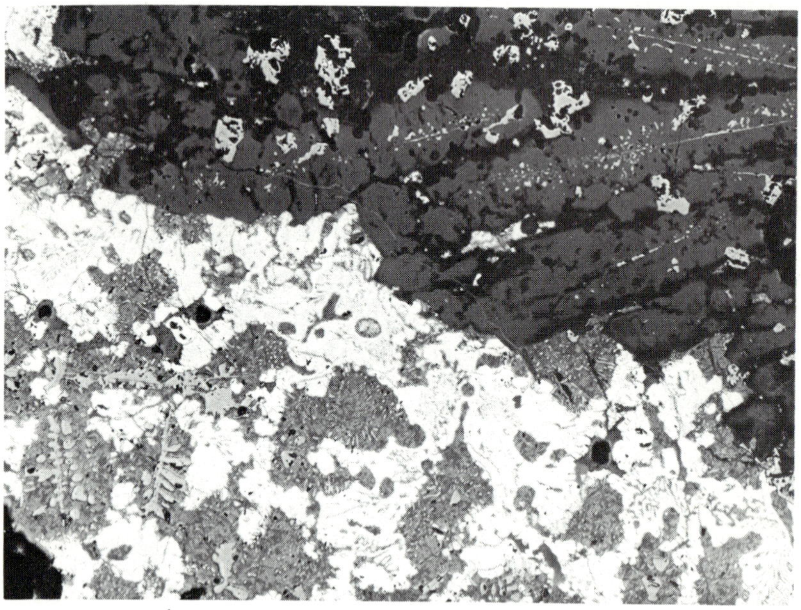

Plate XIe (cf. p.25)

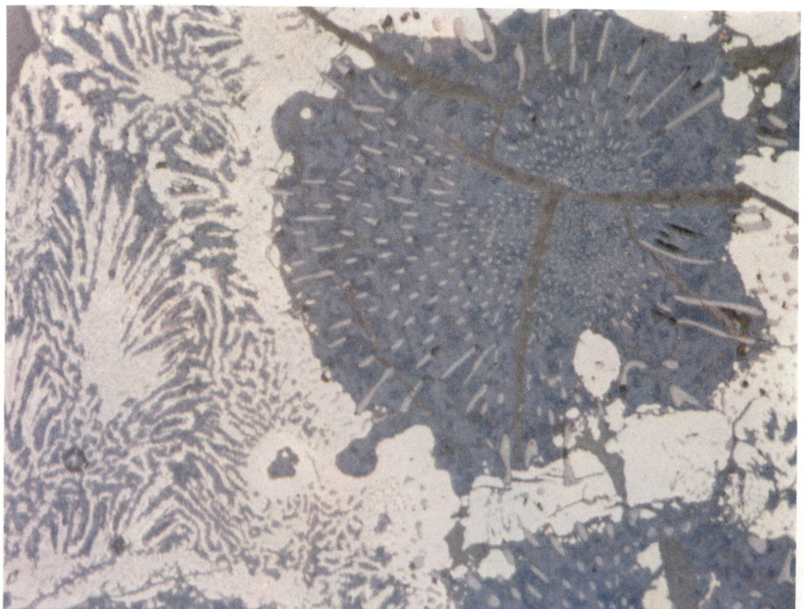

Plate XIf (cf. p.25)

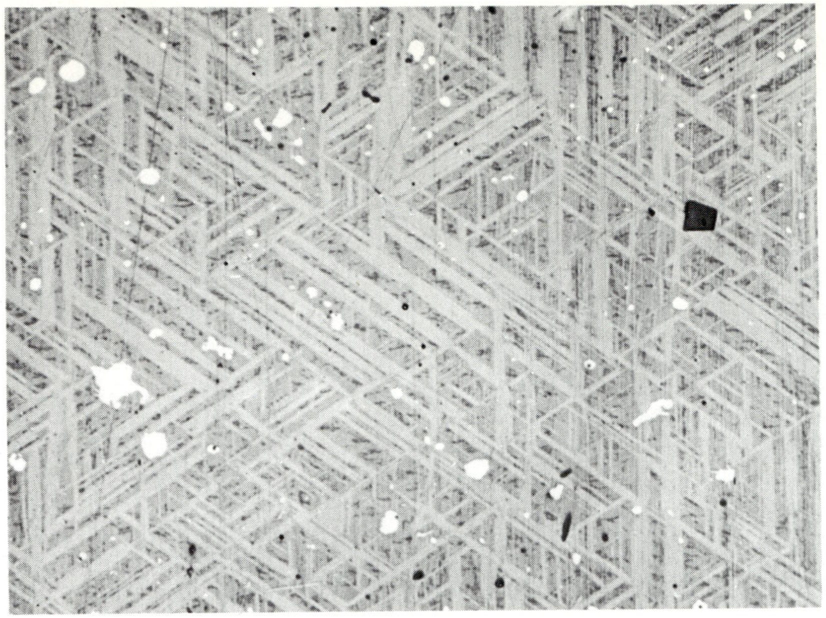

Plate XIIa (cf. p.25)

Plate XIIb (cf. p.25)

Plate XIIIa (cf. p.25)

Plate XIIIb (cf. p.25)

Plate XIIIc (cf. p.26)

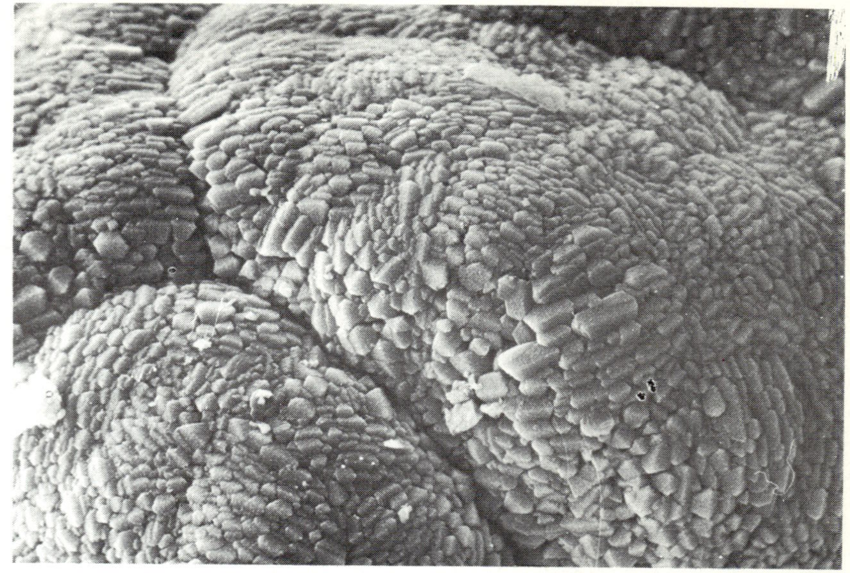

Plate XIIIe (cf. p.26)

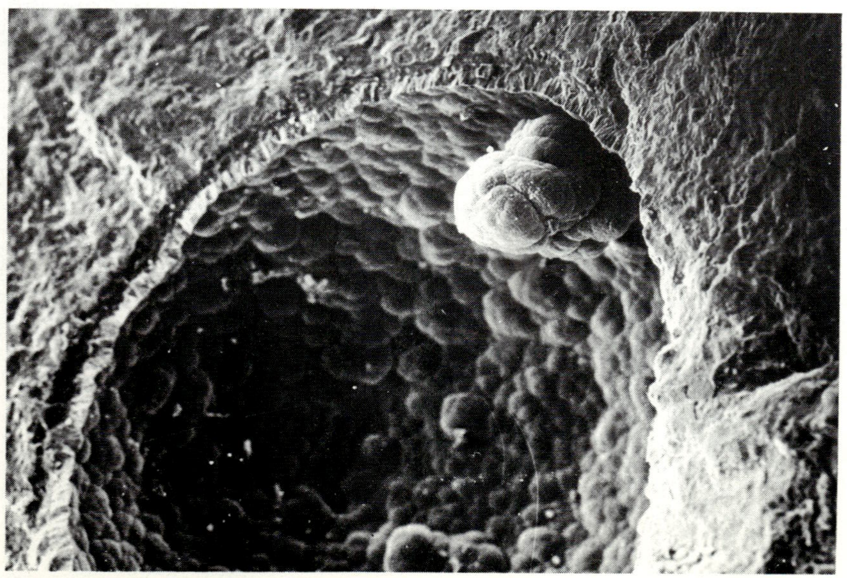

Plate XIIId (cf. p.26)

Plate XIIIf (cf. p.26)

Plate XIVa (cf. p.28)

Plate XIVc (cf. p.28)

Plate XIVb (cf. p.28)

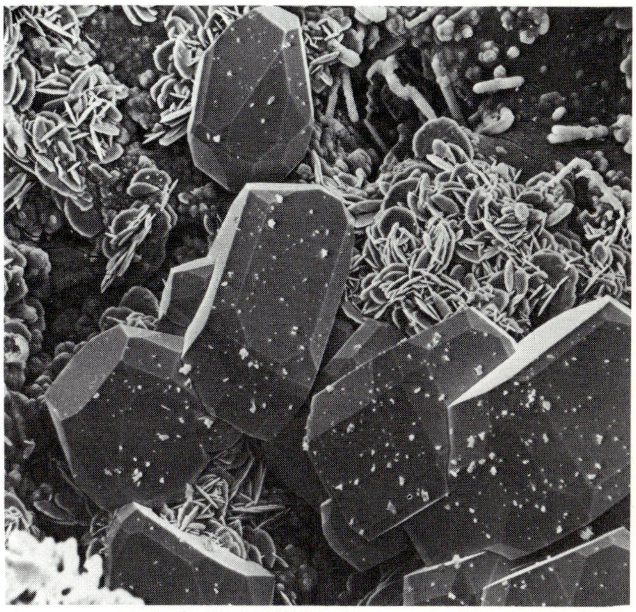

Plate XIVd (cf. p.28)

Plate XIVe (cf. p.28)

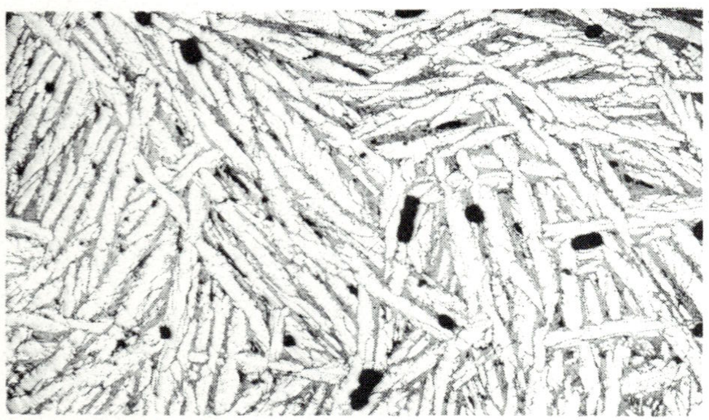
Plate XVa (cf. p.29)

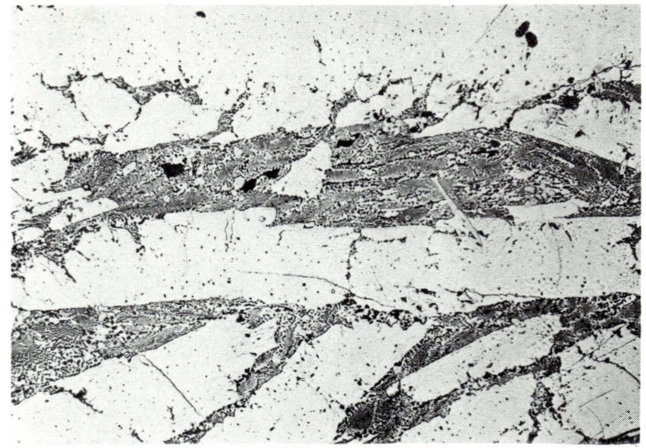
Plate XVb (cf. p.29)

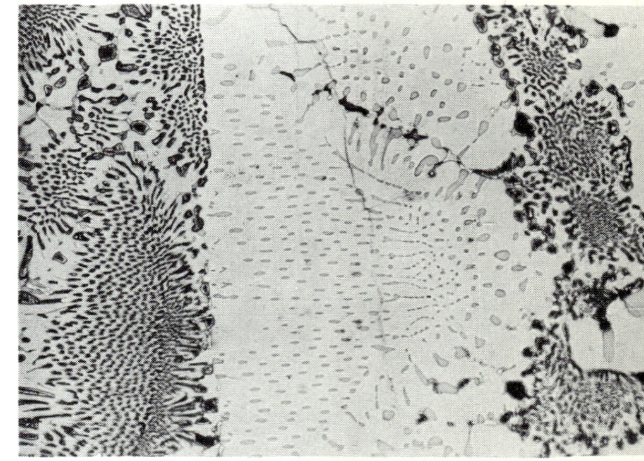
Plate XVc (cf. p.29)

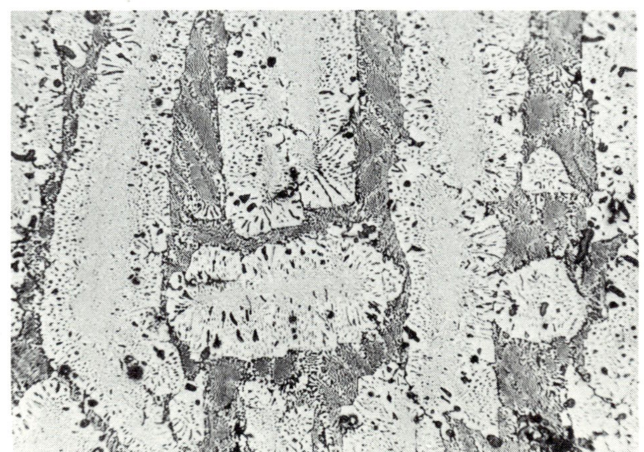
Plate XVd (cf. p.30)

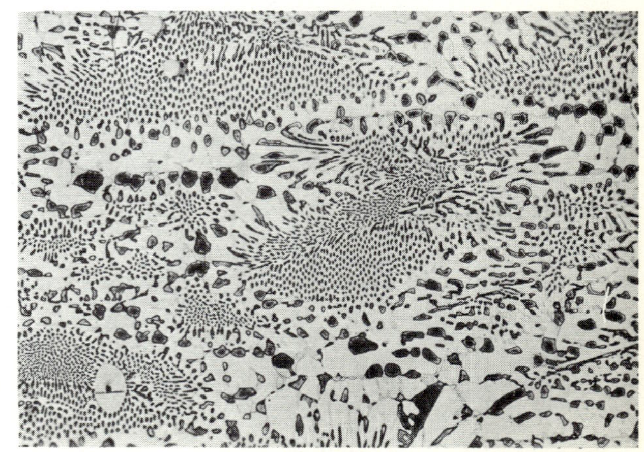
Plate XVe (cf. p.30)

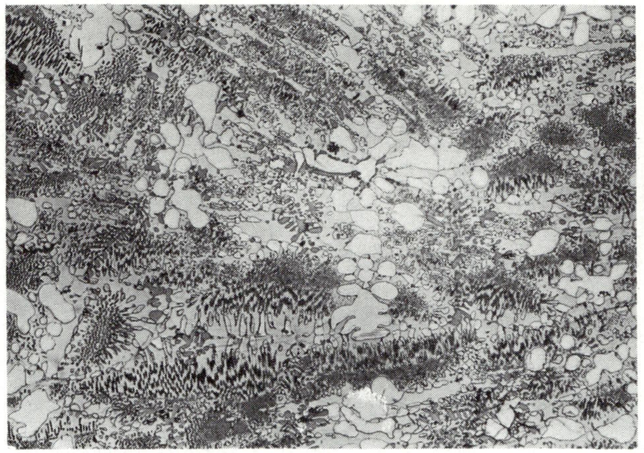

Plate XVf (cf. p.30)

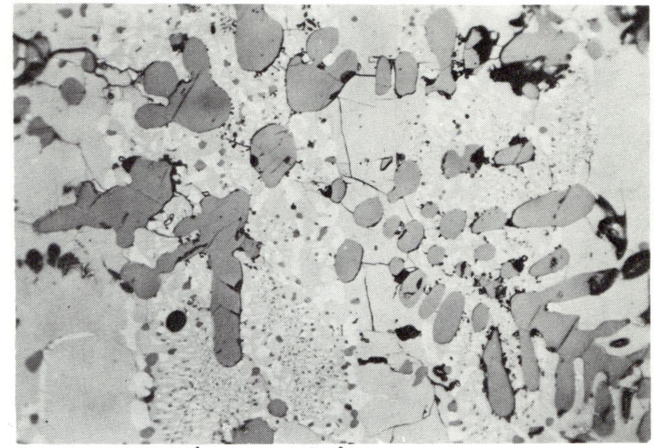

Plate XVh (cf. p.30)

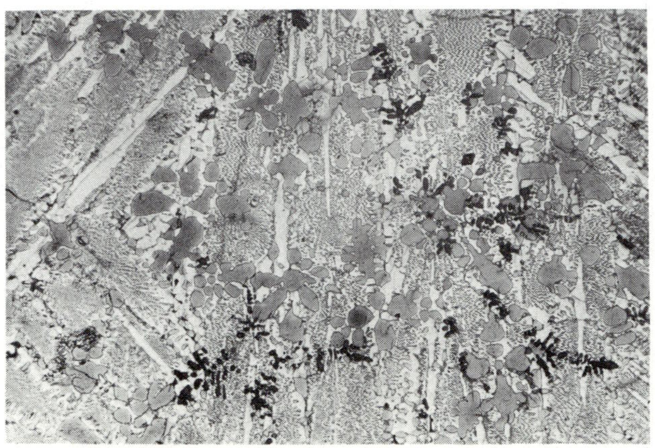

Plate XVg (cf. p.30)

Plate XVi (cf. p.30)

Plate XVIa (cf. p.31)

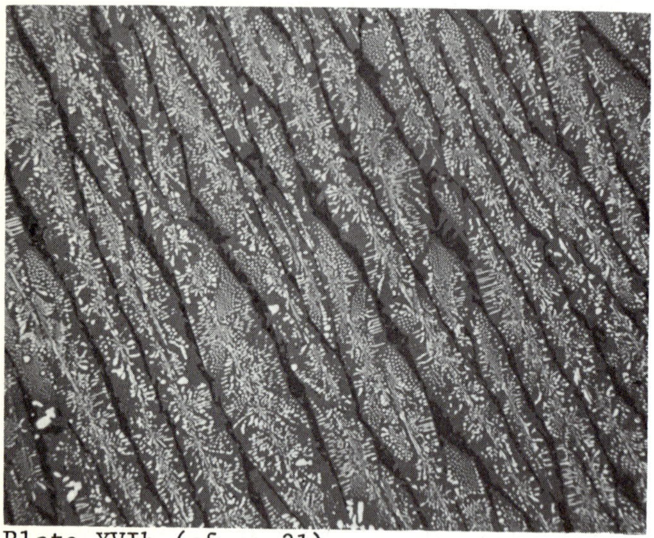

Plate XVIb (cf. p.31)

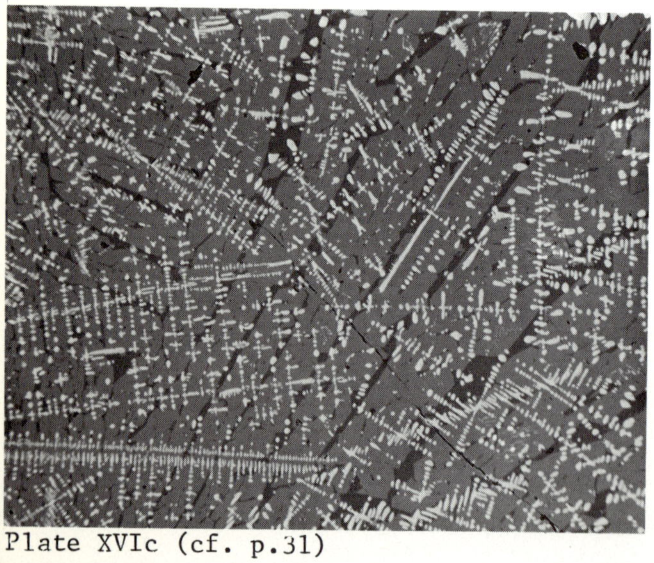

Plate XVIc (cf. p.31)

Plate XVIIa (cf. p.31)

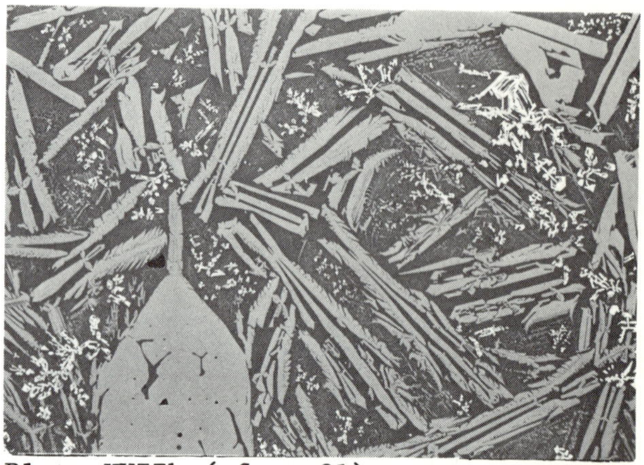

Plate XVIIb (cf. p.31)

Plate XVIIc (cf. p.31)

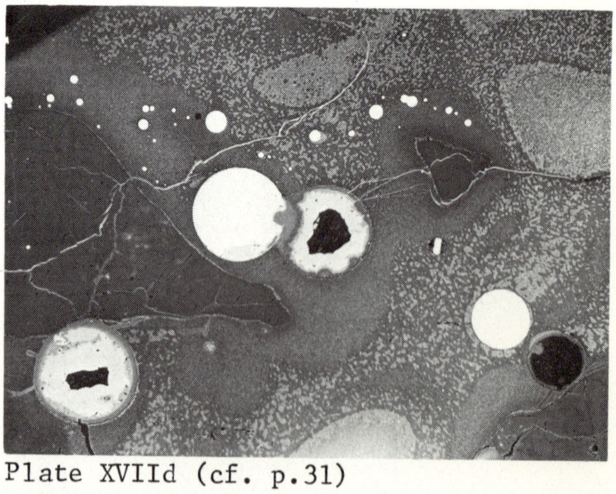

Plate XVIId (cf. p.31)

Plate XVIIIa (cf. p.31)

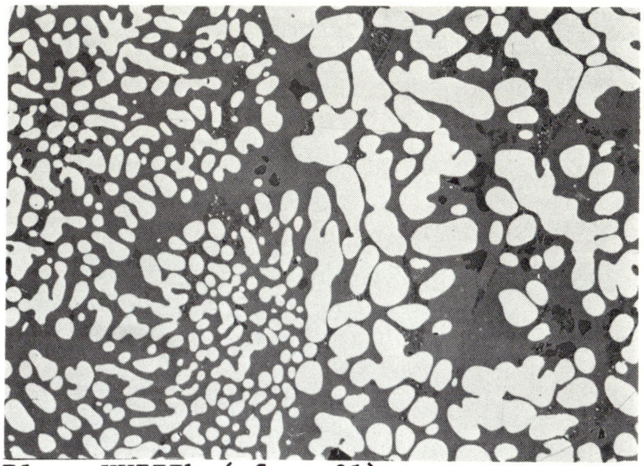

Plate XVIIIb (cf. p.31)

Plate XIXa (cf. p.31)

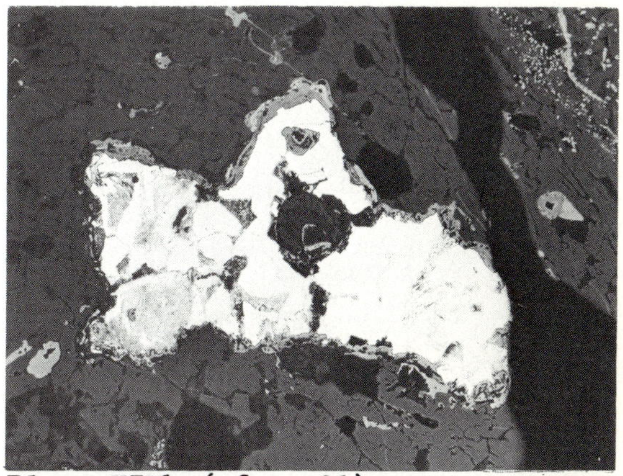

Plate XIXb (cf. p.31)

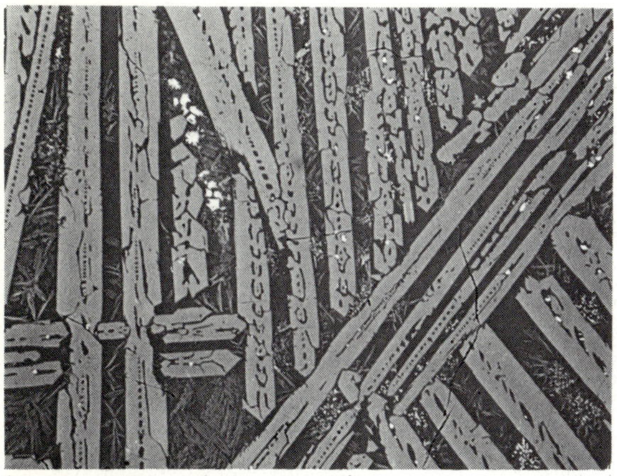

Plate XIXc (cf. p.31)

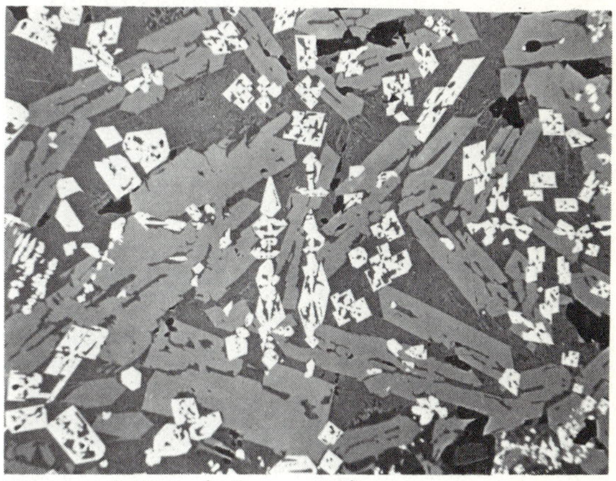

Plate XIXd (cf. p.31)

Plate XIXe (cf. p.31)

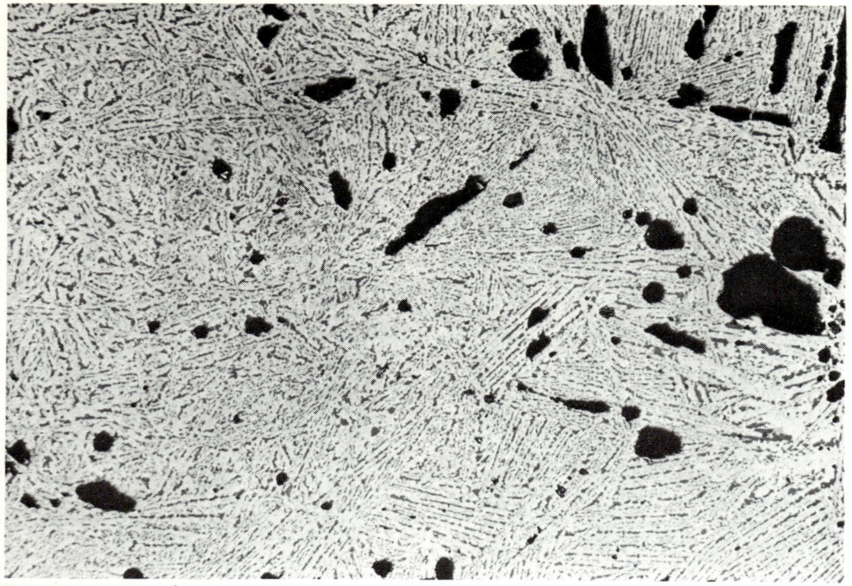
Plate XXa (cf. p.32)

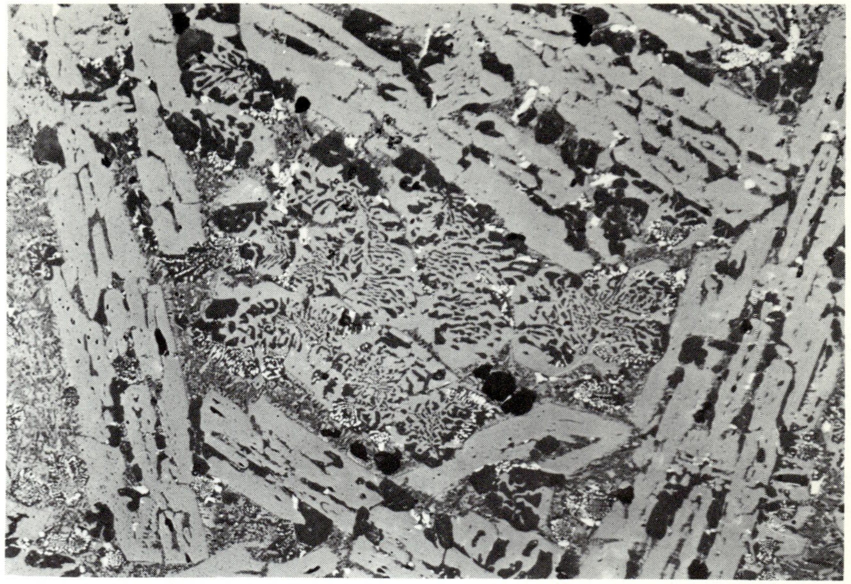
Plate XXb (cf. p.32)

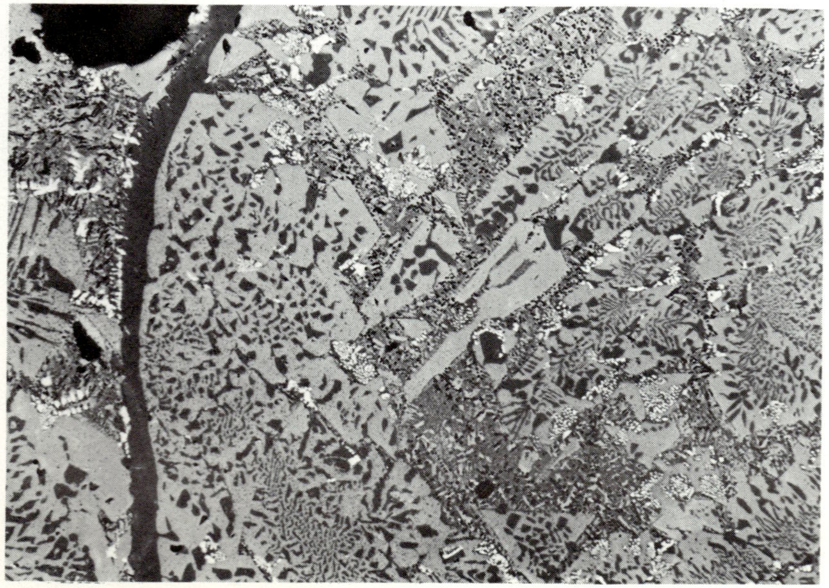
Plate XXc (cf. p.32)

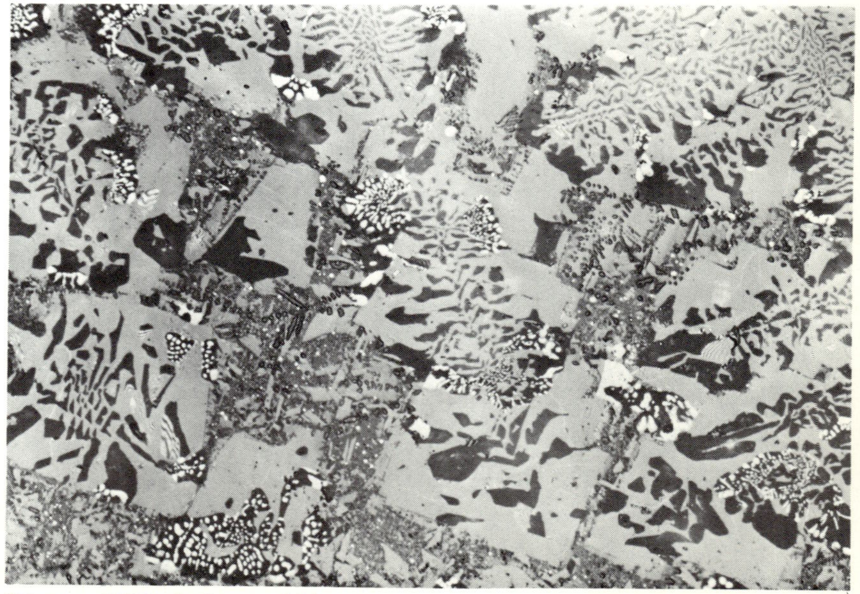
Plate XXd (cf. p.32)

Plate XXIa (cf. p.32)

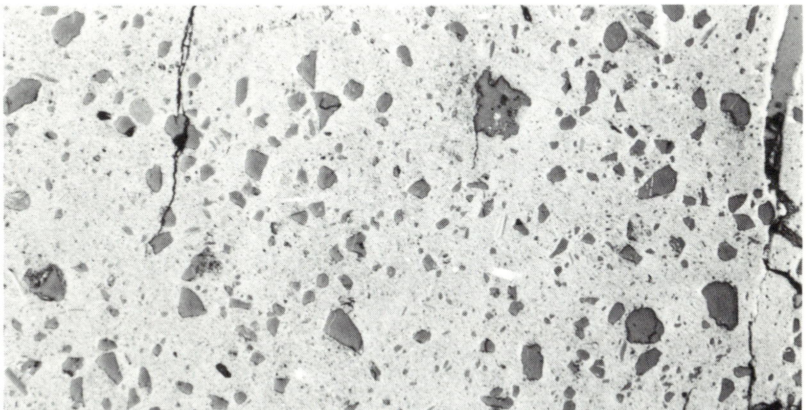

Plate XXIb (cf. p.32)

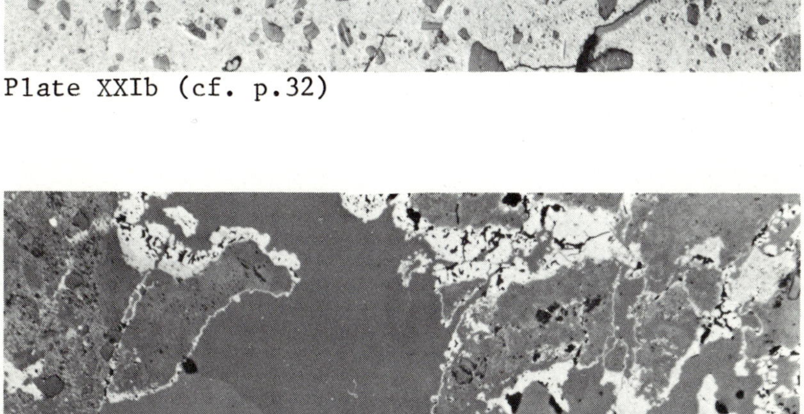

Plate XXIc (cf. p.32)

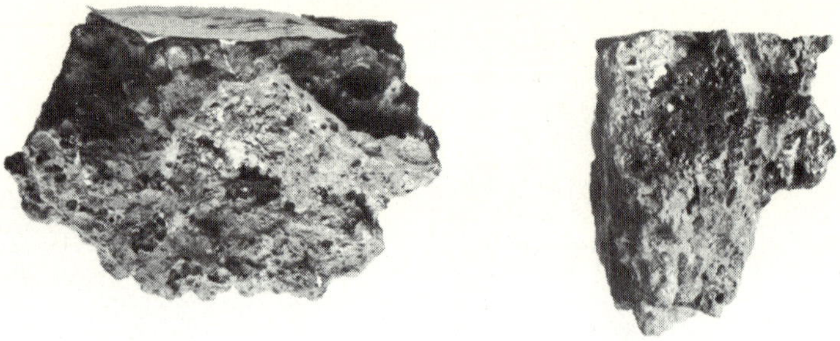

Plate XXIIa (cf. p.32)

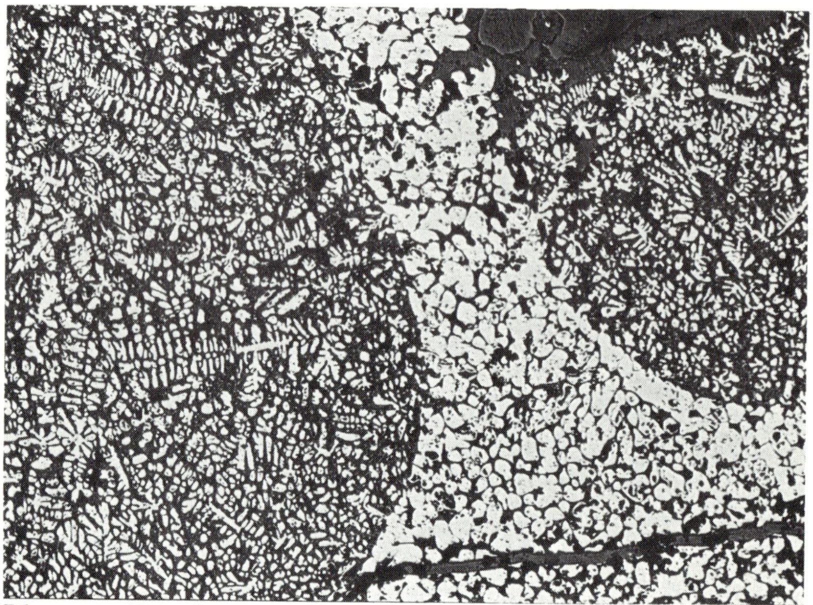

Plate XXIIb (cf. p.32)

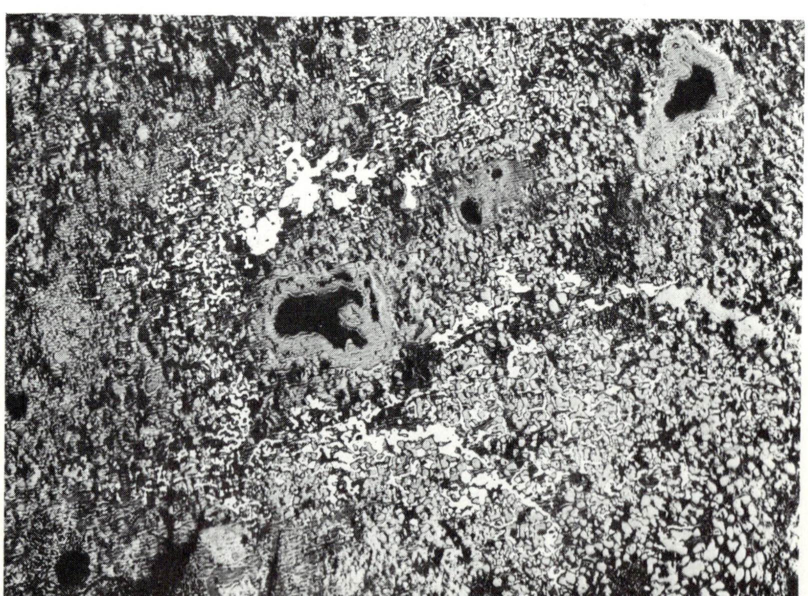

Plate XXIIc (cf. p.32)

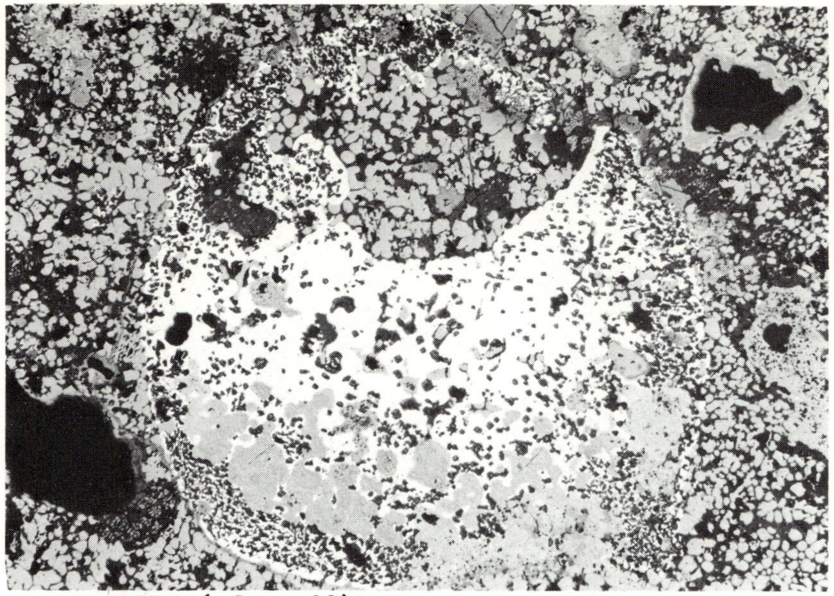

Plate XXIId (cf. p.32)

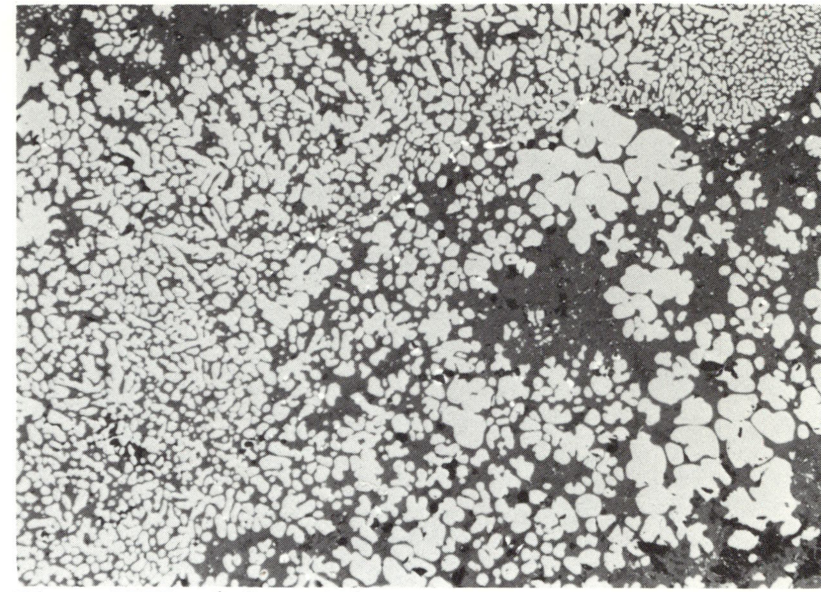

Plate XXIIe (cf. p.32)

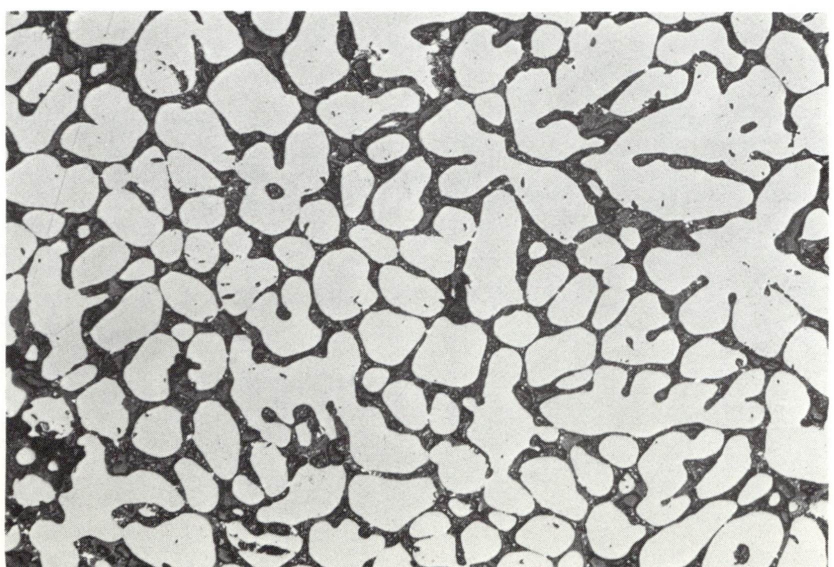

Plate XXIIf (cf. p.32)

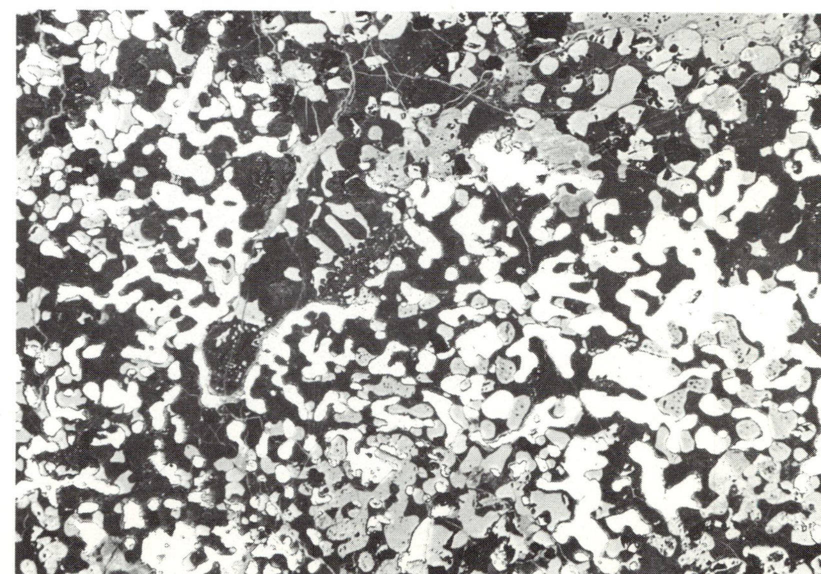

Plate XXIIg (cf. p.32)

Plate XXIIIa (cf. p.32)

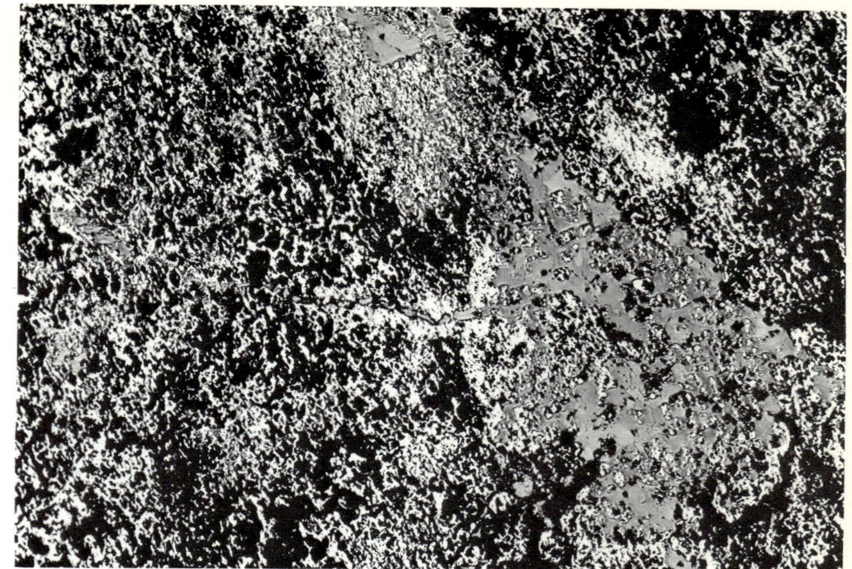

Plate XXIIIb (cf. p.32)

Plate XXIIIc (cf. p.32)

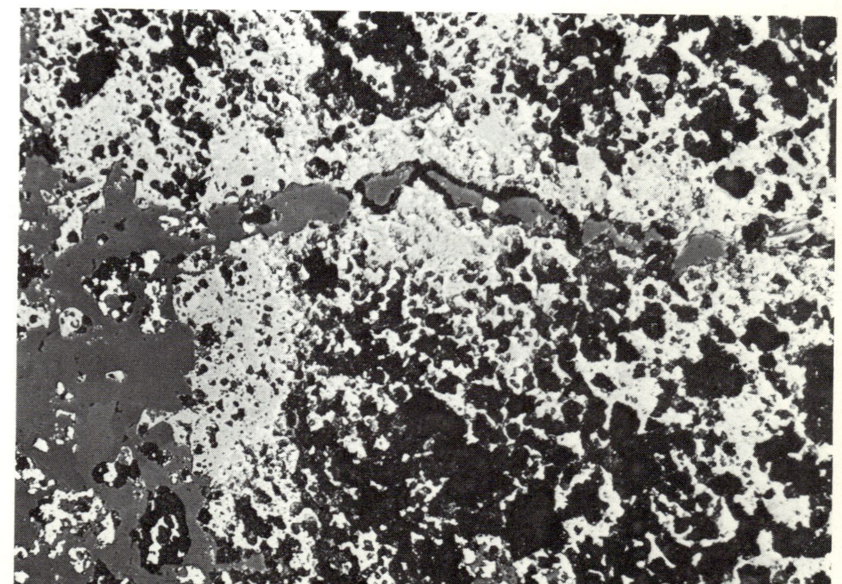

Plate XXIIId (cf. p.32)

Plate XXIVa (cf. p.32)

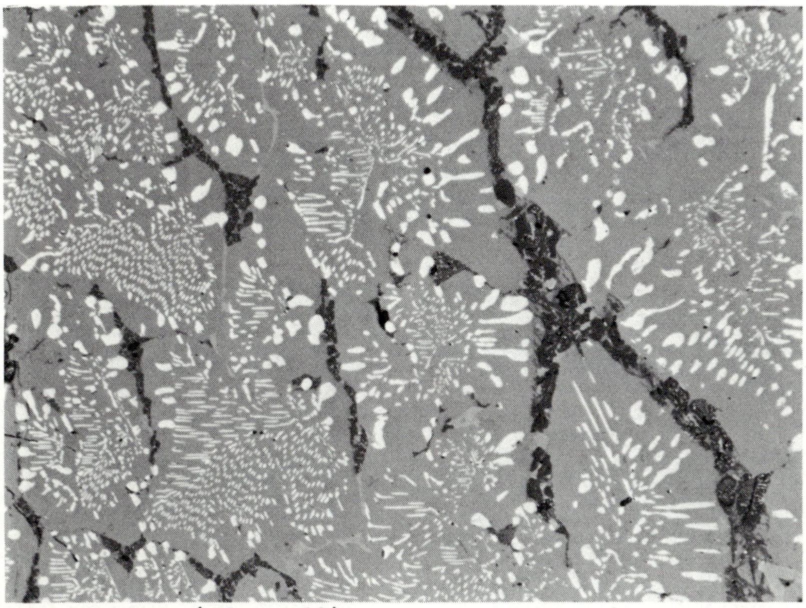

Plate XXIVb (cf. p.32)

Plate XXIVc (cf. p.33)

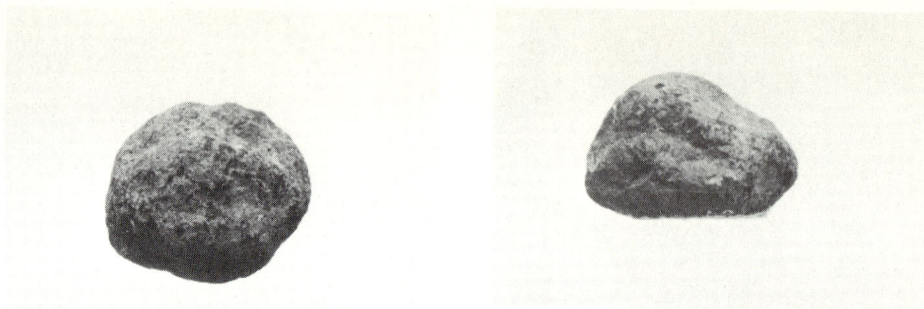

Plate XXVa (cf. p.33)

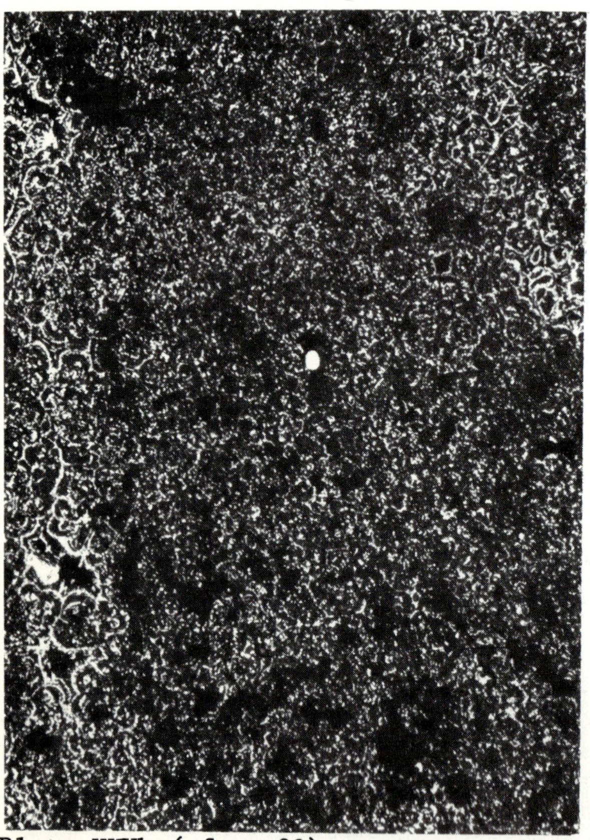

Plate XXVb (cf. p.33)

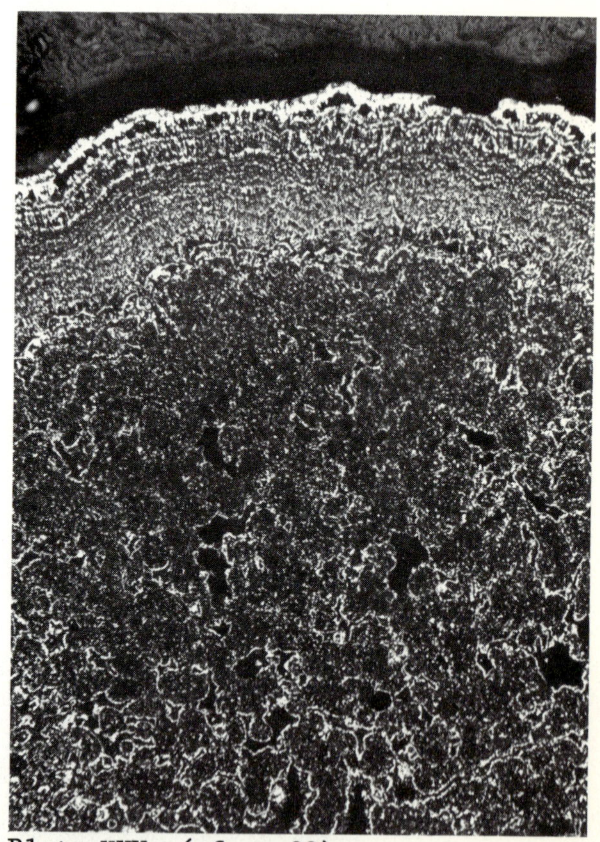

Plate XXVc (cf. p.33)

Plate XXVIa (cf. p.33)

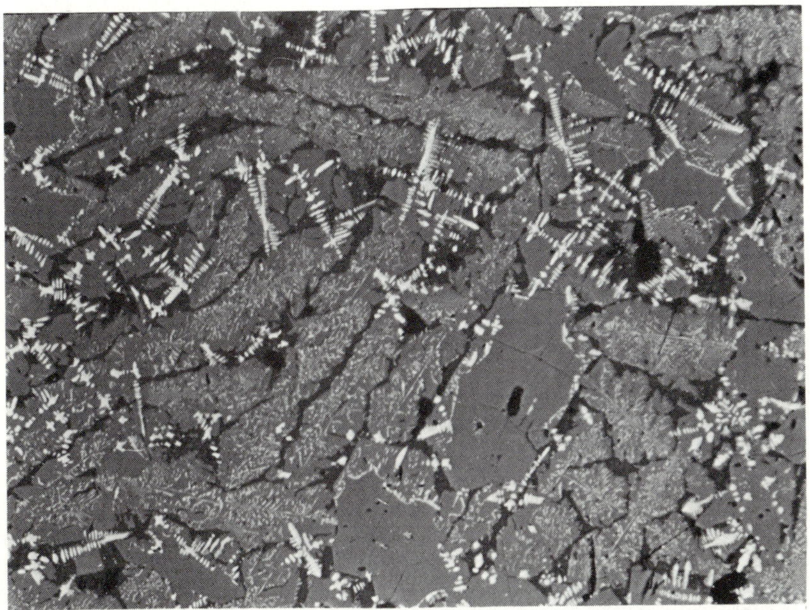

Plate XXVIb (cf. p.33)

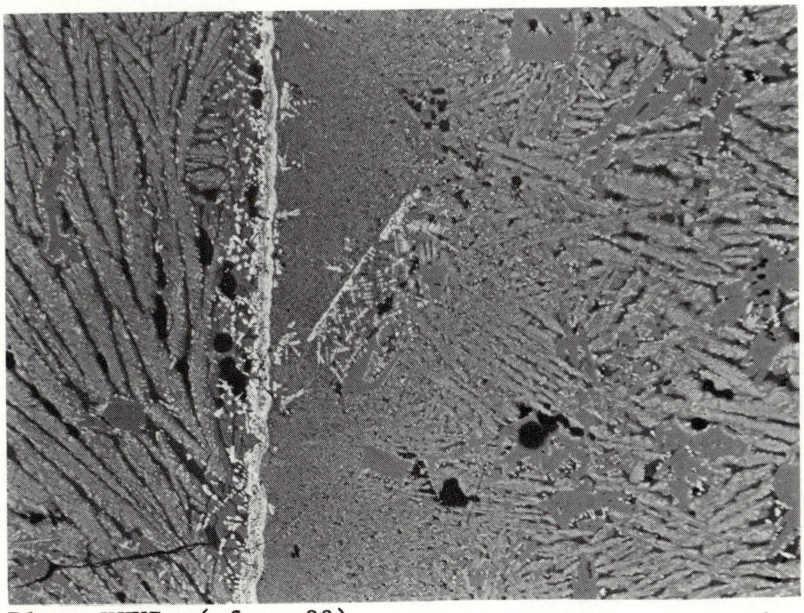

Plate XXVIc (cf. p.33)

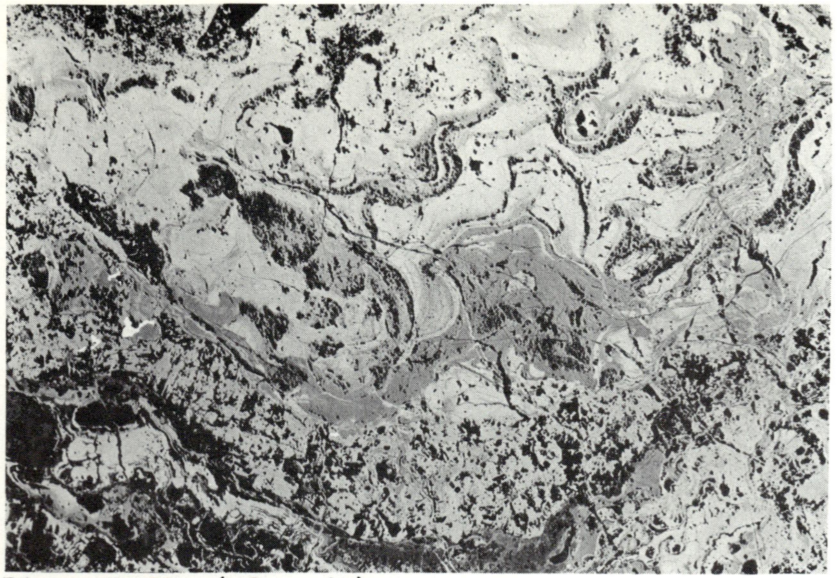
Plate XXVIIa (cf. p.33)

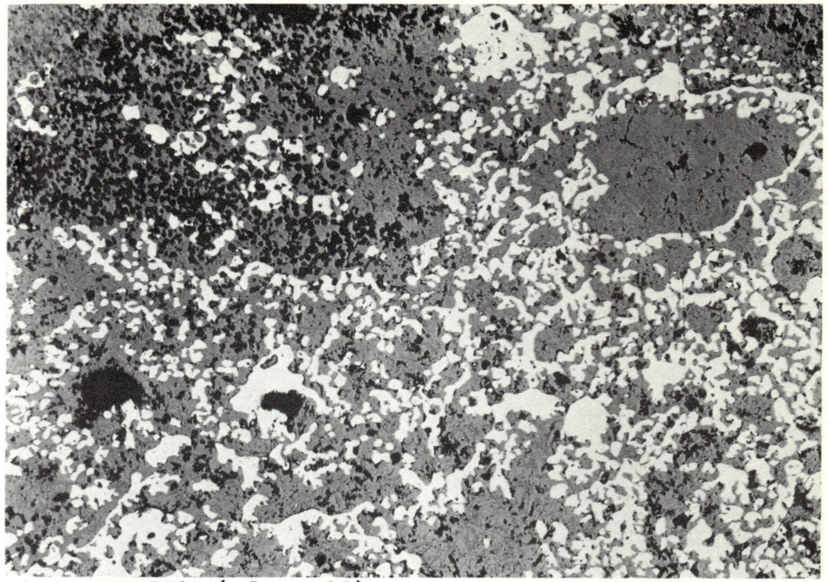
Plate XXVIIb (cf. p.33)

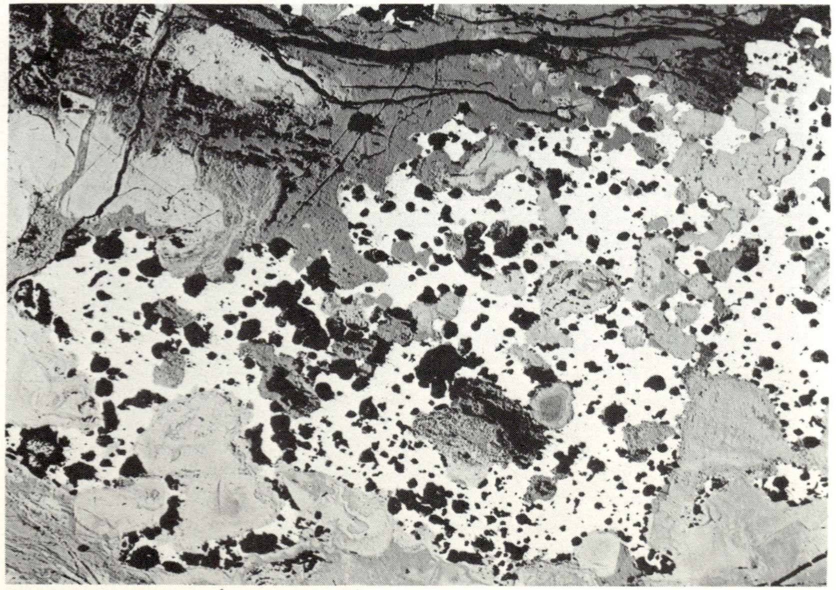
Plate XXVIIc (cf. p.33)

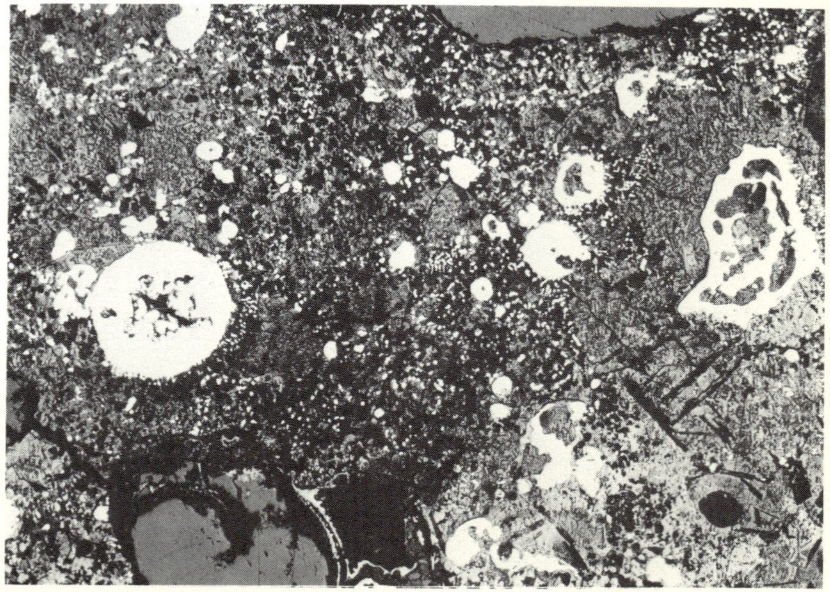
Plate XXVIId (cf. p.33)

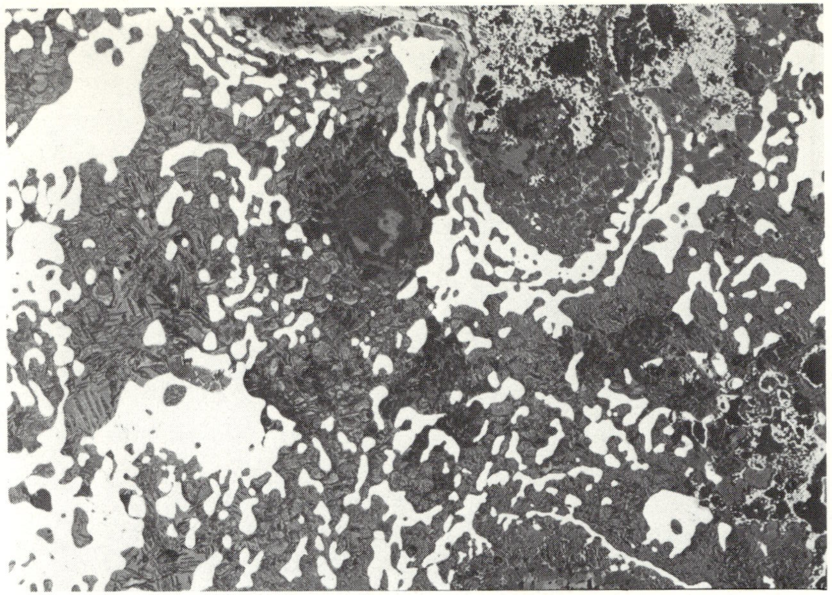
Plate XXVIIe (cf. p.33)

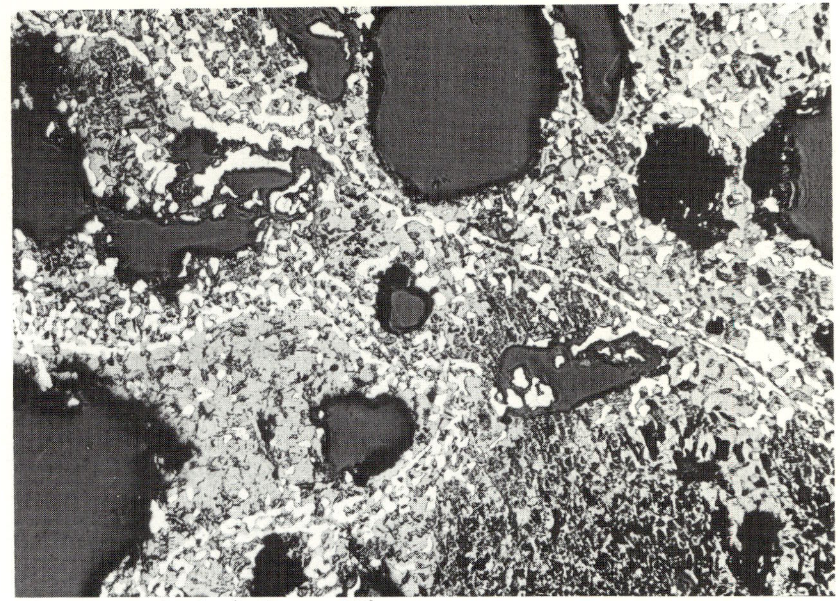
Plate XXVIIf (cf. p.33)

Plate XXVIIg (cf. p.33)

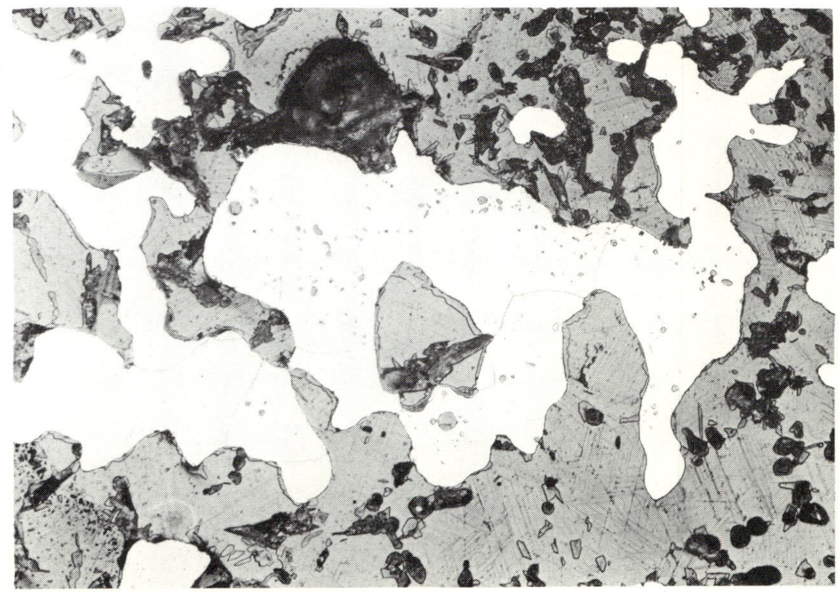
Plate XXVIIh (cf. p.33)